# Introduction to Microscale High-Performance Liquid Chromatography

# Introduction to Microscale High-Performance Liquid Chromatography

**Edited by**

**Daidō Ishii**

**VCH**

e d

Daidō Ishii
Nagoya University
Furo-cho, Chikusa-ku
Nagoya, 464 Japan

**Library of Congress Cataloging-in-Publication Data**

Introduction to micro-scale high-performance
  liquid chromatography

  Includes index.
  1. High performance liquid chromatography.
I. Ishii, Daidō, 1926-
QP519.9.H53I58   1987      543'.0894      87-18908
ISBN 0-89573-309-9

ISBN 0-89573-309-9 VCH Publishers
ISBN 3-527-26636-4 VCH Verlagsgesellschaft

Distributed in North America by:

VCH Publishers, Inc.
220 East 23rd Street, Suite 909
New York, New York 10010

Distributed Worldwide by:

VCH Verlagsgesellschaft mbH
P.O. Box 1260/1280
D-6940 Weinheim
Federal Republic of Germany

# Contributors

**M. Goto**  Research Center for Resource and Energy Conservation, Nagoya University, Chikusa-ku, Nagoya 464, Japan

**K. Hibi**  Japan Spectroscopic Company, Ltd., Hachioji City, Tokyo 192, Japan

**S. Higashidate**  Japan Spectroscopic Company, Ltd., Hachioji City, Tokyo 192, Japan

**D. Ishii**  Department of Applied Chemistry, Faculty of Engineering, Nagoya University, Chikusa-ku, Nagoya 464, Japan

**K. Jinno**  School of Material Science, Toyohashi University of Technology, Toyohashi 440, Japan

**M. Saito**  Japan Spectroscopic Company, Ltd., Hachioji City, Tokyo 192, Japan

**M. Senda**  Japan Spectroscopic Company, Ltd., Hachioji City, Tokyo 192, Japan

**T. Takeuchi**  Department of Applied Chemistry, Faculty of Engineering, Nagoya University, Chikusa-ku, Nagoya 464, Japan

**S. Tsuge**  Department of Synthetic Chemistry, Faculty of Engineering, Nagoya University, Chikusa-ku, Nagoya 464, Japan

**A. Wada**  Japan Spectroscopic Company, Ltd., Hachioji City, Tokyo 192, Japan

# Preface

High-performance liquid chromatography has now come into prominence, due to instrumentation advances in the areas of mechanics, electronics, and computer technology. Much interest and effort in the field of HPLC is now focused on improving selectivity, resolution, speed, and sensitivity. Miniaturization of column dimension in HPLC is considered to be one of the ways to accomplish these improvements.

It has been over a decade since I presented my first paper on *microscale HPLC* at the Tokyo Conference of Applied Spectrometry in 1973. The number of people interested in miniaturization of column dimensions has been increasing since then, however, micro-HPLC is still a little far from the state-of-the art concerning the versatility which conventional HPLC already accomplished.

This book is intended to give introductory and comprehensive information to people who are interested in micro-HPLC and who intend to utilize it for nonspecialized liquid chromatography. This book gives them hints as to the direction of current research along with a directory of applications.

Various terms describing micro-HPLC are found in the literature. In this book, HPLC is classified by the packing state and the column volume (e.g., packed-column HPLC is classified into conventional HPLC, semi-micro-HPLC, and micro-HPLC). It is one of the characteristics of this book that a large portion of the text is dedicated to applications data. I believe this book will educate readers as to what micro-HPLC is, and how it can be utilized.

Finally, I wish to express my indebtedness to the people who have helped me during the preparation of this book. Specifically, I would like to thank Dr. Leslie S. Ettre of the Perkin-Elmer Corporation, who reviewed the original manuscripts and offered many helpful corrections and modifications. I am also grateful to Dr. Edmund H. Immergut of VCH Publishers, Inc. and Mr. Mikio Takahashi of JASCO International Company., Ltd., who first planned for publication of this book. Many thanks go to Miss Masako Matsumae and Mrs. Yoko Amemiya of JASCO for their extensive help with typing the manuscripts.

Daido Ishii

Nagoya University

# Contents

## 4. Detection Systems
### M. Saito, K. Hibi, and M. Goto

## 5. Hyphenated Systems That Employ Microscale Columns
### K. Jinno and S. Tsuge

## 6. Post-Column Derivatization in Microscale HPLC
### M. Senda and S. Higashidate

## 7. Applications of Microscale HPLC
### D. Ishii, T. Takeuchi, and K. Hibi

**Appendixes: List of Available Packing Materials for the Preparation of Packed and Semimicro Columns and Microcolumns**

# Introduction

## D. Ishii and T. Takeuchi

### 1.1. Characteristics of Microscale HPLC

The inner diameter (ID) of the analytical packed columns generally employed in high-performance liquid chromatography (HPLC) is 4–6 mm and commercially available chromatographs have been designed for this column dimension. Column dimension has also been restricted by the cell volume of the detector and fitting techniques. The research on miniaturization of analytical columns in HPLC was initiated by the groups of Horváth,[1] Scott,[2] and Ishii.[3,4] Ishii *et al.* started their work with poly(tetrafluorethylene) (PTFE) tubing of 0.5-mm ID as the column material and they called their research field *micro-HPLC.* The groups of Horváth and Scott employed 1-mm ID stainless-steel columns, calling them packed *microbore columns.* In each case, instrumentation was examined at the laboratory level and the performance of micro-HPLC has been gradually advanced by the progress or improvement of ancillary techniques. Characteristics of micro-HPLC can be naturally generated by employing small-diameter columns. We can expect the following advantages from micro-HPLC:

1. Low consumption of both mobile and stationary phases, which facilitates the use of exotic or expensive phases;
2. Increase in mass sensitivity;
3. Achievement of high resolution with long columns;
4. Applicability of temperature programming;
5. Convenience of selecting the operating conditions;
6. Possibility of coupling with mass spectrometry (MS).

The decrease in column dimensions leads to low consumption of the stationary phase, which facilitates the use of valuable and expensive packing

***D. Ishii and T. Takeuchi*** Department of Applied Chemistry, Faculty of Engineering, Nagoya University, Chikusa-ku, Nagoya 464, Japan.

materials or long columns. Micro-HPLC also facilitates employment of toxic, flammable, or exotic mobile phases, such as low alkanes, carbon dioxide, or deuterated solvents. This allows for the application of micro-HPLC to supercritical fluid chromatography.

Assuming that the same column efficiency is achieved independent of the column diameter, the concentration of solutes eluting from the column is inversely proportional to the square of the inner diameter, i.e., the cross-sectional area of the column. This means that there is a good possibility that micro-HPLC will gain mass sensitivity if a concentration-sensitive detector is used. This increase in mass sensitivity is especially favorable for trace analysis and for analysis of biomedical samples.

In conventional HPLC, it was reported that theoretical plates were not proportional to the column length when several columns were connected in series to produce large theoretical plates. It is one of the advantages of micro-HPLC that reasonable theoretical plates, with respect to column length, can be attained as a result of the decreased multi-path diffusion and effective transfer of heat generated by the pressure drop. The heat generated in the column affects the mass-transfer processes and produces poor column efficiencies. Thus, the temperature in the column should be kept homogeneous. The small heat capacity of microscale columns facilitates application of temperature programming, which is common in gas chromatography (GC), in HPLC. Solvent gradient elution and flow-rate programming have been generally adopted in HPLC. Temperature programming could be applied in the separation of thermally stable compounds, which is commonly achieved by gradient elution.

A large volume of the mobile phase and various types of separation columns are needed to establish the optimum operating conditions for the analysis of new samples, leading to an increased waste of solvents. Micro-HPLC is also convenient for analysis of new samples and a few milliliters of solvent is enough to optimize the mobile phase conditions in micro-HPLC.

Another advantage of micro-HPLC is the possibility of coupling with MS. The lower the flow rate of the mobile phase, the easier direct coupling with MS becomes. This means that the inner diameter of the column should be as small as possible for LC/MS coupling.

Open-tubular (capillary) GC, originally proposed by Golay,[5-7] has been widely utilized since it can achieve larger theoretical plates per unit time and unit pressure drop, due to the good permeability of open-tubular columns, as compared with conventional packed columns. The applicability of open-tubular columns,[8] as well as of packed microcapillary columns,[9] has also been examined in liquid chromatography (LC). These approaches allowed for achievement of large theoretical plate numbers. These columns are distinguished from common packed columns by their packing states.

## 1.2. Classification of Microscale HPLC

Micro-HPLC is classified as such by packing state and column dimension. Micro-HPLC columns can generally be divided into three categories according to their packing states: (1) densely packed columns; (2) loosely packed columns; and (3) open-tubular columns. Figure 1-1 depicts the difference in the packing state of these columns. The first group includes conventional HPLC columns, as well as packed columns, as initiated by the groups of Horváth,[1] Scott,[2] and Ishii.[3] Tsuda and Novotny[9] examined the applicability of the second type of columns in LC and called them packed microcapillary columns. The ratio of the column diameter to the particle diameter is around 2.5 for packed microcapillary columns, which is much smaller than the column-particle diameter ratio for densely packed columns. Permeability of packed microcapillary columns is regarded as falling between the permeabilities of densely packed and open-tubular columns.

Open-tubular columns are quite different in the stationary phase state. In these columns, the stationary phase is present only on the inner surface of the capillary as a thin layer film. Open-tubular columns with the same column dimensions as those used in GC have been employed in LC,[1,10] resulting in poor efficiency, due to the small diffusion speed of LC. Ishii's group[8] employed open-tubular columns with diameters smaller than 60 $\mu$m and showed the applicability of these columns in LC. Characteristics of both packed and open-tubular columns will be discussed in detail in Chapter 3.

Densely packed columns can generally be divided, according to the column diameter, into three categories: conventional HPLC, semi-micro-HPLC, and micro-HPLC. Table 1-1 lists typical column dimensions of these densely packed columns and compares them with those of packed microcapillary and open-tubular columns. Semi-micro-HPLC covers columns with 1 to 2-mm ID and packed microbore columns (initiated by Horváth and Scott) are included in this category. Relative cross-sectional area of the various column types, compared with the 1-mm ID column, are also indicated in Table 1-1. Generally speaking, the column volume of semi-micro-HPLC and micro-HPLC columns is one-tenth and one-hundredth, respectively, of the column volume required for conventional HPLC.

Table 1-2 gives the representative column diameter and flow rate for each category. These calculations are based on conventional HPLC columns

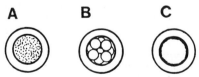

**A**      **B**      **C**

**FIGURE 1-1.** Classification of microscale columns by the packing state. (A) Densely packed column; (B) loosely packed column; (C) open-tubular (capillary) column.

**TABLE 1-1**

Typical Dimensions of HPLC Columns

| Category | ID (mm) | Relative cross-sectional area[a] |
|---|---|---|
| Densely-packed columns: | | |
|   Conventional HPLC | 4–6 | 16–36 |
|   Semi-micro-HPLC | 1–2 | 1–4 |
|   Micro-HPLC | 0.2–0.5 | 0.04–0.25 |
| Loosely packed column (packed microcapillary column) | 0.05–0.2 | 0.0025–0.04 |
| Open-tubular (capillary) column | 0.01–0.06 | 0.0001–0.0036 |

[a]Considering the cross-sectional area of a 1-mm ID column as equal to one.

with 4.6-mm IDs, because they are most frequently employed in LC. Table 1-2 also includes a new category, ultra-micro-HPLC, the column volume of which is defined as one-thousandth of the conventional HPLC column volume. Ultra-micro-HPLC is not specifically discussed in the text because only a few reports have been presented in the literature.[11,12]

Stainless-steel tubing is generally employed as the column tube material in conventional and semi-micro-HPLC, while various types of tubing, such as those made of glass, fused silica, PTFE, and stainless-steel are employed in micro-HPLC. Open-tubular and packed microcapillary columns are prepared from soft glass or Pyrex® glass capillary tubing.

### 1.3. Comparison of Column Efficiencies

The characteristics of open-tubular or packed microcapillary LC are characterized by their ability to achieve high resolution with long columns. Bristow and Knox[13] defined a parameter that compares the column efficiency per unit time and unit pressure drop. This is the separation impedance ($E$) expressed as:

$$E = \frac{t_0}{N} \times \frac{P}{N} \times \frac{1}{\eta} \tag{1}$$

**TABLE 1-2**

Representative Column Diameters and Flow Rates in Various Types of HPLC

| Category | ID (mm) | Flow rate ($\mu$ l/min) |
|---|---|---|
| Conventional HPLC | 4.6 | 1000 |
| Semi-micro-HPLC | 1.5 | 100 |
| Micro-HPLC | 0.46 | 10 |
| Ultra-micro-HPLC | 0.15 | 1 |

**TABLE 1-3**
Comparison of the Dimension of High-Resolution Columns[a]

| Category | ID ($\mu$m) | Length (m) | Particle diameter ($\mu$m) | $dc/d_p^b$ | $\phi$ |
|---|---|---|---|---|---|
| Densely packed columns | 250–1000 | 1–14 | 5–30 | 50–200 | 500–1000 |
| Loosely packed columns | 50–200 | 20–100 | 10–100 | 2–8 | 100–300 |
| Open-tubular (capillary) columns | 10–60 | 20–40 | — | — | 32 |

[a]Adapted from Knox.[14]
[b]$d_c$, column diameter; $d_p$ particle diameter.

where $t_0$ is the elution time of an unretained solute, $N$ is the number of theoretical plates, $\Delta P$ is the pressure drop along the column, and $\eta$ is the viscosity of the mobile phase. This parameter is convenient in comparing column efficiencies obtained using different operating conditions or different types of columns. $E$ is dimensionless and can be expressed by using dimensionless parameters as follows:[14]

$$E = h^2\phi \qquad (2)$$

where $h$ is the reduced plate height and $\phi$ is the column resistance parameter. The reduced plate height in open-tubular LC is defined as the ratio of the plate height to the column diameter.[14] The smaller the separation impedance, the better the column efficiency.

Table 1-3 compares the dimension of high-resolution columns and the column resistance factor.[14]

## References

1. Horváth, C.G.; Preiss, B.A.; Lipsky, S.R. *Anal. Chem.* **1967, 39,** 1422.
2. Scott, R.P.W.; Kucera, P. *J. Chromatogr.* **1976, 125,** 251.
3. Ishii, D. *Jasco Report* **1974,** 11 (no. 6).
4. Ishii, D.; Asai, K.; Hibi, K.; Jonokuchi, T.; Nagaya, M. *J. Chromatogr.* **1977, 144,** 157.
5. Golay, M.J.E. *Anal. Chem.* **1957, 29,** 928.
6. Golay, M.J.E. *Anal. Chem.* **1957, 180,** 435.
7. Golay, M.J.E. In "Gas Chromatography 1958", Desty, D.H., Ed.; Butterworths: London, 1958; p 36.
8. Hibi, K.; Ishii, D.; Fujishima, I.; Takeuchi, T.; Nakanishi, T. *J. High Resolut. Chromatogr. Chromatogr. Commun.* **1978, 1,** 21.
9. Tsuda, T.; Novotny, M. *Anal.Chem.* **1978, 50,** 271.
10. Nota, G.; Marino, G.; Buonocore, V.; Ballio, A. *J. Chromatogr.* **1970, 46,** 103.
11. Takeuchi, T.; Ishii, D. *J. Chromatogr.* **1980, 190,** 150.
12. Takeuchi, T.; Ishii, D. *J. Chromatogr.* **1981, 218,** 199.
13. Bristow, P.A.; Knox, J.H. *Chromatographia* **1977, 10,** 279.
14. Knox, J.H. *J. Chromatogr. Sci.* **1980, 18,** 453.

# Instrumental Requirements in Microscale HPLC

## M. Saito, K. Hibi, D. Ishii, and T. Takeuchi

### 2.1. Introduction

The history of microscale HPLC goes back almost as far as that of modern ordinary-scale HPLC. It started in 1967, When Horváth and co-workers investigated the operating parameters of nucleotide separation on a small-bore column having an internal diameter of 1 mm and a length of about 2 m with a comparably large column volume.[1,2] A new HPLC, *called micro-HPLC,* originated in 1973 when Ishii and co-workers succeeded in separating polynuclear aromatic hydrocarbons with increased sensitivity by using a homemade micro-HPLC system that incorporated an extremely small volume PTFE column (0.5-mm ID × 150-mm length) that was packed with a pellicular material.[3,4,5] In 1976, Scott and Kucera demonstrated a highly efficient separation of alkylbenzenes on a 1-mm ID × 10-m long column, which was used with a modified ultraviolet (UV) detector that had a 2.4-$\mu$l flow cell.[6,7] In the same year, the first micro-HPLC system,[8] the model FAM-ILIC-100 (FAst MIcro LIquid Chromatograph; shown in Fig. 2-1) became commercially available from JASCO.* The system was equipped with a screw-driven glass syringe pump and a variable-wavelength UV detector, with a 0.3-$\mu$l micro flow cell. However, until recently, the practical use of micro-HPLC has been generally neglected by most chromatographers.

Early HPLC columns were of 1.6- to 3-mm ID and 500- to 1000-mm long, packed with 30-$\mu$m pellicular material. When totally porous microparticulate packing materials, with particle diameters of 5 to 10 $\mu$m, were intro-

*Japan Spectroscopic Company, Ltd., 2967-5 Ishikawa-cho, Hachioji City, Tokyo 192, Japan.

**M. Saito and K. Hibi** Japan Spectroscopic Company, Ltd. Hachioji City, Tokyo 192, Japan. **D. Ishii and T. Takeuchi** Department of Applied Chemistry, Faculty of Engineering, Nagoya University, Chikusa-ku, Nagoya 464, Japan.

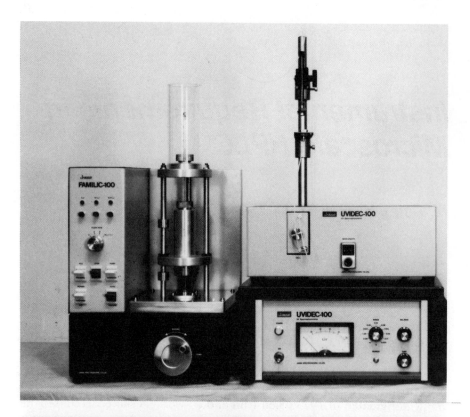

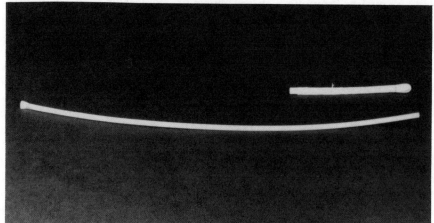

**FIGURE 2-1.** JASCO FAMILIC-100 and its PTFE column. The first commercial micro-HPLC system. (Courtesy of JASCO.)

duced, the column inner diameter was increased to 5–6 mm. At that time most chromatographers believed that small-bore columns performed poorly. The main reasons for preferring large-bore columns may have included the limited technology for packing the small-bore columns with 5- to 10-$\mu$m particles and the existence, at that time, of detectors that were inadequate for the narrow peaks commonly eluted by small-bore columns. Such detectors were designed to be used primarily with 2.1-mm ID × 500-mm long columns that have 500 theoretical plates. In other words, they were intended to detect peaks corresponding to a mobile phase volume of 200 $\mu$l or greater. On the other hand, microparticulate packing material can yield 10,000 plates per 250-mm column length. This means that the peak width for an unretained component is only 24 $\mu$l if the column dimension is 2.1-mm ID × 250-mm length. Such a column is obviously not compatible with a conventional 8-$\mu$l detector cell.

For this reason, when the microparticulate materials appeared, the entire HPLC system should have been redesigned. Instead, the column internal diameter was increased.[9] A 4.6-mm ID × 250-mm long column results in a mobile phase volume of about 100 $\mu$l for the unretained peak. Peaks of this width could be detected with an 8-$\mu$l cell without any significant loss of column efficiency, but the compromise results in a significant loss in sensitivity.

Chromatography is a separation process but, at the same time, it is a dilution process. The dilution coefficient is proportional to the column length and the square of the column diameter, and inversely proportional to the square root of the plate number of the column. This means that the concentrations of sample components eluted by a 4.6-mm ID column are almost as low as the values obtained from an older pellicular column,[10] even though the number of theoretical plates is 20 times higher than that of the older column. Thus, if a chromatographer intends to obtain optimal performance from a modern microparticulate packing material, he must carefully reexamine fundamental factors such as the peak volume, maximum peak concentration, flow rate, injection volume, detector cell volume, and so on.

## 2.2. Fundamental Factors in Microscale HPLC

### 2.2.1. Peak Volume

Peak volume is the most fundamental factor to be taken into account when designing an HPLC system in which a small-volume column is used, because it determines most of the instrumental requirements. Assuming that the solute concentration distribution is Gaussian, the volume $V_p$ of a peak eluting from a column having a theoretical plate number $N$ corresponding to four standard deviations ($4\sigma$) of the distribution curve can be expressed as

$$V_p = 4\sigma = \frac{4V_R}{\sqrt{N}} = \frac{4V_o(1 + k')}{\sqrt{N}} \tag{1}$$

**TABLE 2-1**

Peak Volume Corresponding to the Unretained Peak Obtained on Various
Columns and the Relative Peak Concentration Obtained on These Columns

| ID in mm | Length (L) in mm | Theoretical plate number (N) | Peak volume ($V_p$) in $\mu$l | Relative peak concentration |
|---|---|---|---|---|
| 2.1 | 500 | 500 | 200 | 0.5 |
| 4.6 | 250 | 10,000 | 100 | 1 |
| 1.5 | 250 | 10,000 | 10 | 10 |
| 0.5 | 250 | 10,000 | 1 | 100 |

where $V_R$ is the retention volume, $V_o$ is the vacant volume of the column
(the retention volume of an unretained peak), and $k'$ is the capacity factor.
The column vacant volume ($V_o$) can be expressed from the column diameter
($d_c$), column porosity ($\epsilon$), and column length ($L$) as

$$V_o = \frac{\pi d_c^2 \epsilon L}{4} \tag{2}$$

Equation 1 then becomes

$$V_p = 4\sigma = \frac{\pi d_c^2 \epsilon L (1 + k')}{\sqrt{N}} \tag{3}$$

Thus, the peak volume is proportional to the square of the column diameter
and to the column length, and inversely proportional to the square root of
the number of theoretical plates. Table 2-1 lists the peak volumes that correspond to the unretained peaks for various columns.

### 2.2.2. Maximum Peak Concentration

In the chromatographic separation process, a sample solute injected onto a
column disperses in a certain amount of mobile phase solvent when it elutes
from a column. More quantitatively, about 95% of the sample solute dis-
tributes in a $4\sigma$ peak volume ($V_p$).

The total sample solute is calculated by integrating the concentration
distribution curve, i.e., the peak area $A$, and is represented as

$$A = \sqrt{2\pi}\, \sigma C_{max} \tag{4}$$

Where $C_{max}$ is the maximum peak concentration and $\sigma$ is the standard deviation of the concentration distribution, assumed as Gaussian, in terms of solvent volume. Therefore, the maximum peak concentration ($C_{max}$) of a given
sample mass ($m_s$) can be expressed, replacing $\sigma$ with $V_p/4$ in equation 4 as

$$C_{max} = \frac{2\sqrt{2}\, m_s}{\sqrt{\pi}\, f_R V_p} \cong \frac{1.6 m_s}{f_R V_p} \tag{5}$$

or

$$C_{max} = \frac{2\sqrt{2}\, m_s\sqrt{N}}{f_R \pi^{3/2} d_c^2 \epsilon L(1 + k')} \tag{5a}$$

where $f_R$ is the detector response factor for the sample mass ($m_s$; $m_s = f_R A$). Equation 5a shows that the maximum peak concentration $C_{max}$ is proportional to the square root of the theoretical plate number ($N$), and inversely proportional to the column length ($L$) and the square of the column diameter ($d_c^2$). Using equation 5a, the relative peak concentration of various columns having different dimensions and efficiencies can be calculated. Such values are given in Table 2-1. As can be seen, reduction of the column dimensions, especially the diameter, drastically increases the maximum peak concentration. This means that the detection sensitivity for the same sample mass could be increased in this way. This is one of the most important advantages of using a small-volume column.

### 2.2.3. Flow Rate

The flow rate of the mobile phase is one of the most important factors to be taken into account in order to properly downscale the HPLC system. The linear velocity of the mobile phase ($u$) is

$$u = L/t_o \tag{6}$$

where $L$ is the column length and $t_o$ is the retention time of an unretained solute. The volumetric flow rate ($F$) is expressed by using the column vacant volume ($V_o$), as

$$F = V_o/t_o \tag{7}$$

Substituting $V_o/F$ for $t_o$ and using equations 2 and 6, one can express the relationship between the linear velocity ($u$) and the volumetric flow rate ($F$) as

$$u = \frac{4F}{\pi d_c^2 \epsilon} \tag{8}$$

Thus, in order to maintain the same linear velocity for columns of different diameters, the volumetric flow rate ($F$) should be decreased in proportion to the square of the ratio of column diameters.

Figure 2-2 shows the relationship between the flow rates and column diameters for common linear velocities used in normal analysis. Table 2-2 lists peak volumes, peak heights (peak maximum concentrations), flow rates, and solvent consumption, as calculated for columns of various dimensions under the same chromatographic conditions and with the same theoretical plate number and linear velocity. As shown, the 1.5- and 0.5-mm ID columns give ten and one hundred times higher peak concentrations, respectively, than a conventional 4.6-mm ID column with the same length.

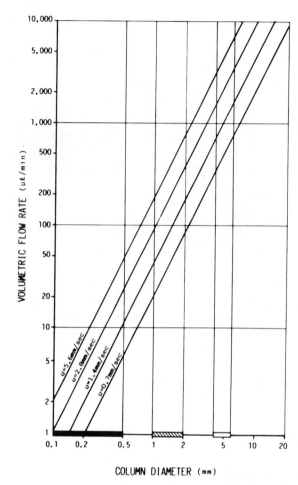

COLUMN DIAMETER (mm)

**FIGURE 2-2.** Volumetric flow rates of columns with various diameters for the same linear velocity. □, Diameter range of conventional columns; ▨, diameter range of semi-microcolumns; ■, diameter range of microcolumns. Column porosity ($\epsilon = 0.7$) for all columns.

## TABLE 2-2
Comparisons of the Chromatographic Parameters in Conventional, Semi-microcolumns, and Microcolumns[a]

| | | Column type (length, mm; ID, mm) | | |
|---|---|---|---|---|
| | | Conventional (250; 4.6) | Semi-microcolumn (250; 1.5) | Microcolumn (250; 0.5) |
| Linear mobile phase velocity ($u$) | mm/sec | 1.4 | 1.4 | 1.4 |
| Flow rate ($F$) | ml/min | 1.0 | 0.1 | 0.01 |
| Peak volume ($V_p$) | $\mu$l | 116 | 12 | 1.4 |
| Peak height ($C_{max}$) | % | 1.2 | 12 | 100 |
| Solvent consumption per 7 hour operation | ml | 420 | 42 | 4.2 |

[a]Column porosity ($\epsilon$), 0.7; capacity factor ($k'$), 0; and theoretical plate number ($N$), 10,000.

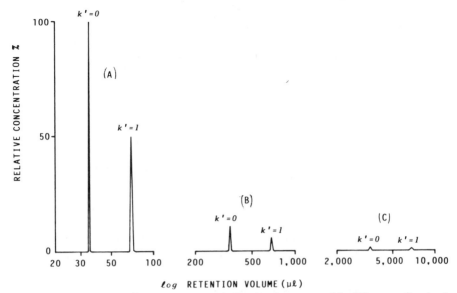

**FIGURE 2-3.** Schematic chromatograms calculated and represented for (A) conventional column (4.6-mm ID × 250-mm length); (B) a 1.5-mm ID × 250-mm long column having one-tenth the volume of a conventional column; and (C) a 0.5-mm ID × 250-mm long column having one-hundredth the volume of a conventional column.

While the solvent consumption of the former columns is reduced by the same factor compared with the conventional column. Figure 2-3 is the schematic representation of chromatograms for the same sample mass obtained on the conventional (4.6-mm ID) column and on the columns with 1.5- and 0.5-mm ID.

We can conclude from this discussion that the typical dimensions of a conventional column (4.6-mm ID × 250-mm length) are by no means adequate for a microparticulate packing material; a column of smaller dimensions is more suitable for such a column packing.

## 2.3. Instrumental Requirements for HPLC

### 2.3.1. Microscale Columns

Various names have been used to characterize the small-volume columns. Microbore column seems to be the most widely accepted. However, columns should not be categorized by the bore (internal diameter) but by the column volume, since columns having the same vacant volume give the same peak volume irrespective of the bore. Bearing this in mind, we should call a column of one hundredth or less the volume of a conventional column as *microcolumn* and a column of about one-tenth the volume as *semi-micro-column,* distinguishing them from each other.

**TABLE 2-3**
Maximum Extracolumn Peak Volumes for Peaks Eluted by Various Columns[a]

| | Column type (length, mm; ID, mm) | | |
|---|---|---|---|
| | Conventional (250; 4.6) | Semi-microcolumn (250; 1.5) | Microcolumn (250; 0.5) |
| Maximum extracolumn peak volume ($V_{ex}$, $\mu$l)[b] | 53 | 5.5 | 0.6 |
| Peak volume ($V_p$) | 116 | 12 | 1.4 |

[a]Column porosity ($\epsilon$), 0.7; capacity factor ($k'$), 0; and theoretical plate number ($N$), 10,000.
[b]10% increase in peak volume is allowed.


Detailed description of microcolumns and semi-microcolumns is given in Chapter 3. Their essential features, as compared to a conventional column, are:

1. Peak volume, volumetric flow rate, and solvent consumption are reduced by a factor of 100 for microcolumns and 10 for semi-microcolumns.
2. Peak concentration is increased by a factor of 100 for microcolumns and 10 for semi-microcolumns.

In addition, semi-microcolumns have the advantage that they are now available from a number of column suppliers, and that they are compatible with most conventional pumps. Suitable detectors (mainly UV) and injectors are also commercially available.

In order to obtain good performance from microcolumns, it is very important to have an HPLC system with very low dead volume, so that the peak broadening outside the column does not destroy the resolution achieved within the column. Pumps and injectors for use with microcolumns have recently become commercially available (see section 3-2).

### 2.3.2 Extracolumn Peak Broadening

Extracolumn peak broadening has been investigated by many researchers.[1,3,4,7,9,11–24] This section will just give a brief review.

The variance (the square of the standard deviation) of the observed peak ($\sigma_{p(ob)}^2$) can be expressed as the sum of the peak variance caused only by the contribution of the column ($\sigma_p^2$) and all the contributions to extracolumn peak broadening ($\sigma_{ex}^2$)[11]. This is expressed as

$$\sigma_{p(ob)}^2 = \sigma_p^2 + \sigma_{ex}^2 \qquad (9)$$

Since the peak volume is four times the standard deviation ($\sigma$), equation 9 can be rewritten as

$$V_{p(ob)}^2 = V_p^2 + V_{ex}^2 \qquad (10)$$

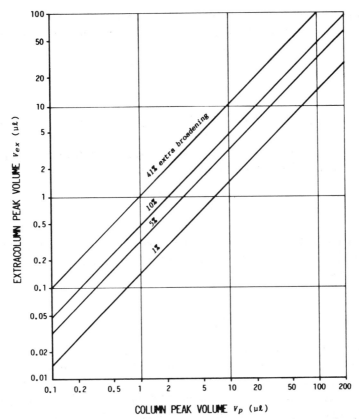

**FIGURE 2-4.** Relationship between column peak volumes and extracolumn peak volumes for various degrees of extracolumn broadening.

where $V_p$ is the peak volume obtained only by column contribution and $V_{ex}$ is the extracolumn peak volume corresponding to the contributions of the injector, detector cell, and connecting tubing. Dividing equation 2-10 by $V_p^2$ produces

$$(V_{p(ob)}/V_p)^2 = 1 + (V_{ex}/V_p)^2 \qquad (11)$$

Therefore, if the observed peak is allowed to have a volume 10% greater than the column peak volume ($V_p$), the extracolumn peak volume should be about one half (46%) of the column peak volume. Table 2-3 lists column peak volumes and maximum extracolumn peak volumes for various types of columns. Figure 2-4 shows the relationship between the column and extracolumn peak volumes for various degrees of extracolumn broadening.

### 2.3.3. Major Causes of Peak Broadening

There are several causes of peak broadening in an HPLC system. The Poiseuille flow dispersion, the diffusion chamber effect, and the mixing chamber effect are the major causes of extracolumn peak broadening.

In the Poiseuille flow, the velocity profile in a tube with circular cross-section is described by the Hagen–Poiseuille equation and becomes parabolic as

$$u(r) = \frac{\Delta P}{4\eta l_t} (r_o^2 - r^2)$$  (12)

where $u(r)$ is the flow velocity as a function of the distance ($r$) from the tube center, $r_o$ is the internal radius of the tube, $l_t$ is the tube length, $\Delta P$ is the pressure drop along the tube, and $\eta$ is the viscosity. In the center of the tube ($r = 0$), the velocity becomes maximum and is calculated as

$$u_{max} = \frac{\Delta P\, r_o^2}{4\eta l_t}$$  (13)

while on the wall of the tube ($r = r_o$), it becomes minimum to zero. The average velocity can be calculated by integrating equation 12 and dividing by the cross-sectional area.

$$u_{av} = \frac{\Delta P\, r_o^2}{8\eta l_t} = \frac{u_{max}}{2}$$  (14)

Therefore, the sample solute in the center penetrates at twice the average flow velocity, on the other hand, the portion of the sample solute on the wall surface would stay forever on the wall, if there were not radial mass transfer of the sample solute. This mass transfer takes place by only molecular diffusion; this is very small in a liquid. As a conclusion, the *left-behind* portion of the sample solute on the wall slowly diffuses toward the place where the concentration is lower and it will be swept by the flow. In this way, the advancing portion and the remaining portion of the sample solute cause the dispersion. Figure 2-5 shows the schematic representation of the Poiseuille flow dispersion. It was investigated theoretically and experimentally by Taylor,[25,26] confirmed by many researchers, and intensively described in the literature.[4,11,12,17,18,23,27] It should be noted that, as will be explained in Section 2.3.7, the peak variance of the Poiseuille flow dispersion is proportional to the flow rate.

The diffusion and mixing chamber effects are inevitably involved when the flow passes through tubing, which has the abrupt diameter changes associated with dead volumes. The injector, the detector cell, and the connecting tubes and fittings cause this type of peak broadening. The diffusion chamber effect is similar to the Poiseuille flow dispersion to such an extent that the remaining portion of the sample solute diffuses slowly. However, the flow velocity profile is not simple and is dependent on the geometry of the path; its contribution is therefore generally not as calculable as is the Poiseuille flow dispersion in long straight tubing (see Eq. 2-25). Sternberg[11] showed that diffusion and mixing chamber behavior is functionally similar, but with

**A**

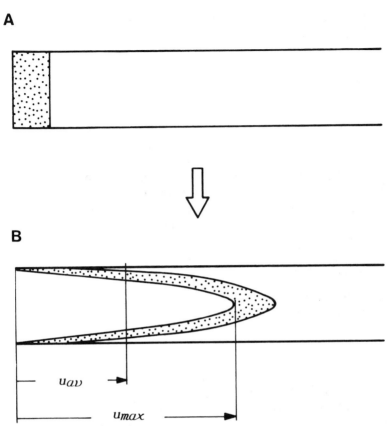

**B**

$u_{av}$

$u_{max}$

**FIGURE 2-5.** Schematic representation of Poiseuille flow dispersion. The sample solute initially distributed as a complete rectangular zone (A) disperses as it moves in a tube due to the parabolic profile of the flow velocity (B).

the time constant diffusion-controlled in the first case and flow rate-controlled in the second case. The time constant of the concentration decay of the sample solute, due to the diffusion chamber effect, is inversely proportional to the coefficient of molecular diffusion of the sample solute, while, due to mixing chamber effect, it is inversely proportional to the flow rate. Peak broadening that results from those effects is exponential in nature and leads to asymmetrical, tailing peaks.[12] As the flow rate becomes higher, eddies in the flow, rather than molecular diffusion, take a major part in the radial mass transfer of the sample solute. When the concentration becomes uniform in the chamber by eddy-mixing, the chamber can be treated as a perfect mixing chamber. The time variance $\sigma_t^2$ of the mixing chamber is expressed as

$$\sigma_t^2 = (V_m/F)^2 \tag{15}$$

where $V_m$ is the mixing chamber volume and $F$ is the flow rate. The variance $\sigma_v^2$ can be represented in terms of mobile phase volume as

$$\sigma_v^2 = V_m^2 \tag{16}$$

and becomes independent of the flow rate.

In the actual hydraulics of an HPLC system, these phenomena are combined in a complex way. Especially in micro-HPLC, the flow velocity in the tubing is generally lower than that in conventional HPLC, thus the diffusion and mixing chamber effects can contribute more significantly than in conventional HPLC. Accordingly, even the smallest dead volume must be eliminated and the flow path should be as simple as possible.

In the next section the contributions of the injection volume, the cell volume, and the time constant of the detector and connecting tubing will be examined and the instrumental requirements for a microscale-HPLC system will be calculated.

### 2.3.4. Injection Volume

When a sample having a volume of $V_{inj}$ is introduced as a complete rectangular zone (perfect plug), the sample peak variance $\sigma_s^2$ can be expressed as

$$\sigma_s^2 = V_{inj}^2/12 \tag{17}$$

If 5% extra peak-broadening is allowed, the allowable sample injection volume can be calculated (by inserting equation 17 into equation 11) to be about one third (28%) of the column peak volume. However, perfect plug injection cannot be achieved by an actual injector; therefore, the value should be less than one third depending on the injection methods, i.e., the initial sample solute distribution.[11,14,28–30]

Several sample introduction methods have been reported for micro-HPLC, including stopped-flow,[3,4,31] split-flow,[32,33] heart-cut,[34] and miniaturized valve injection[7,35,36] techniques. Some of them are suitable for injecting nanoliter level sample volume, but they are less practical for routine analysis. This discussion will cover miniaturized valve injectors, which are suitable for practical applications. These injectors are now commercially available from several sources.

An injector that has a simple and smooth flow path is recommended as the sample solute is not left behind at any portion of the path. The diffusion and mixing that occurs in an injector that has a complicated flow path often cause severe tailing, especially at lower flow rates of less than 100 $\mu$l/min. An injector with a simple and smooth path was reported by Takeuchi and Ishii[35]; its configuration is as shown in Fig. 2-6.[36]

Figure 2-7 shows chromatograms that were obtained on a semi-micro-column (1.5 mm ID $\times$ 250 mm length) with 1-$\mu$l injection volume, using a conventional loop injector (JASCO VL-614) and a microvalve injector with the above configuration (JASCO ML-425). Chromatogram A shows fairly

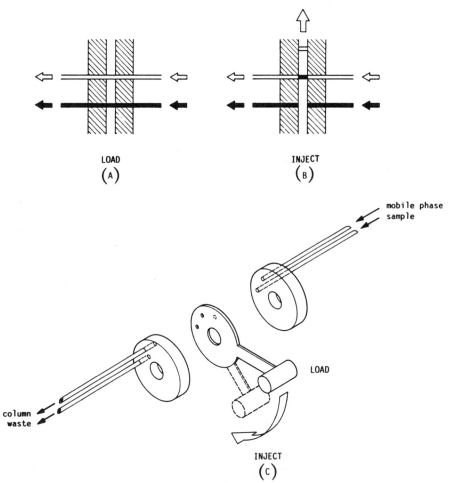

**FIGURE 2-6.** (A) and (B): Operating principle of the internal-loop valve injector with a simple rotor disc. (C) Injection of sample solution into the column by changing the disc position from LOAD to INJECT. □, Mobile phase; ■, sample.

good resolution, while chromatogram B shows extremely poor resolution that was caused by the injector. Table 2-4 compares the theoretical plate numbers for the same column, obtained from the chromatograms shown in Fig. 2-6.

Table 2-5 lists maximum injection volumes for various peaks, allowing for 5% increase in the peak volumes. It should be noted that the maximum injection volumes listed in Table 2-5 are valid only when there is no adsorption of the solute in the stationary phase, as in size exclusion chromatography. If the sample is dissolved in a solvent weaker than the mobile phase, the sample would then be enriched at the top of the column bed and elution

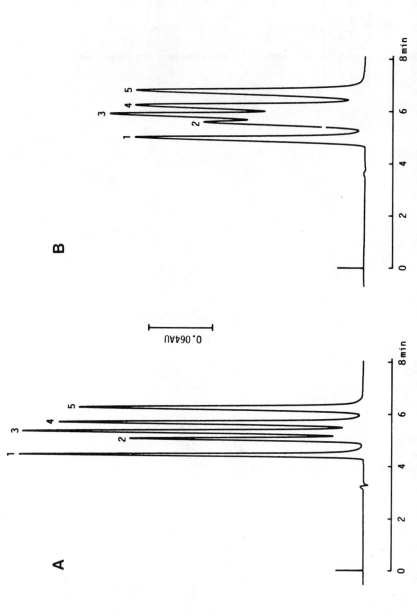

**FIGURE 2-7.** Chromatograms obtained by using different injectors. Chromatogram (A) was obtained using a conventional injector. Chromatogram (B) was obtained using a microvalve injector while chromatogram (B) was obtained using a microvalve injector while chromatogram (B) was obtained using a conventional injector. Chromatographic conditions: column, μ S-FinePak SIL C₁₈ (1.5-mm ID × 250-mm length). Mobile phase: acetonitrile/water (90/10). Flow rate: 100 μl/min. Injection volume: 1 μl. Detector: UVIDEC-100-V with a 1-μl cell monitored at 250 nm. Sample: 1, benzene (1.0%); 2, naphthalene (0.05%); 3, biphenyl (0.01%); 4, fluorene (0.01%); 5, anthracene (0.001%). Sample solvent was acetonitrile.

**TABLE 2-4**

Effect of the Injector on the Efficiency of a Microscale Column[a]

| | Number of theoretical plates | |
|---|---|---|
| | Benzene | Anthracene |
| Microvalve injector | 9,600 | 10,000 |
| Conventional injector | 3,700 | 5,200 |

[a]Theoretical plate numbers were calculated from the chromatograms shown in Fig. 2-7, by measuring the width at 0.607 of the peak height. Injected sample volume was 1 $\mu$l. Other chromatographic conditions are given in the caption for Fig. 2-7.

takes place later. Therefore, a sample volume more than the above calculated value can be applied without significant broadening. On the other hand, if the sample solute is dissolved in a solvent stronger than the mobile phase, the calculated maximum sample volume should be strictly respected. Furthermore, in the extreme case where a much stronger solvent is used, the sample solvent might take part in the mobile phase, resulting in shorter retention times than expected. For this reason, the sample should be dissolved in the same solvent as in the mobile phase or, preferably, in a weaker solvent. Figure 2-8 shows the relationship between the observed peak volumes and injection volumes for the same sample amount dissolved in different solvents.

Figure 2-9 shows the effect of the injection volume on column efficiency in micro-HPLC using a 0.25-mm ID $\times$ 98-mm long column.[37] The sample is dissolved in a stronger solvent. For solutes having $k' \geq 2$, a slight deterioration in column efficiency is observed even with 0.05-$\mu$l injection. It should also be noted that, as the peak volume is proportional to the column volume, the allowable injection volume is dependent on both the column diameter and length.

Figure 2-10 demonstrates[35] the good reproducibility of valve injection and the performance of microcolumns. Four columns were prepared and each column was subjected to three or four injections of the sample. It is remarkable that the 0.6% relative standard deviation for the peak heights of successive 15 measurements was obtained using one of the above columns, even though the injection volume is as small as 0.02$\mu$l. An ML-422 micro-

**TABLE 2-5**

Maximum Injection Volumes for Various Peak Volumes[a]

| Maximum injection volume ($\mu$l) | Peak volumes ($\mu$l) |
|---|---|
| 0.3 | 1 |
| 3 | 10 |
| 30 | 100 |

[a]Perfect plug injection is assumed and a 5% increase in the peak volume is allowed.

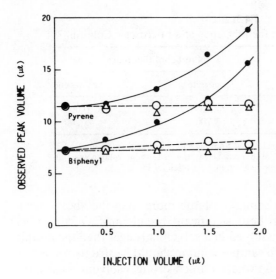

INJECTION VOLUME (µℓ)

**FIGURE 2-8.** Observed peak volumes and injection volumes for the same sample amount dissolved in different solvents. ●, Dissolved in acetonitrile; ○, dissolved in acetonitrile/water (70/30); △, dissolved in acetonitrile/water (50/50). Column: 0.5-mm ID × 150-mm long PTFE tube packed with ODS SC-01.

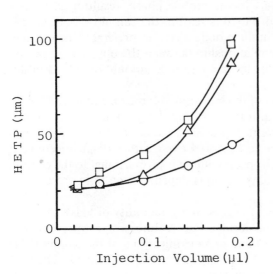

**FIGURE 2-9.** The influence of the injection volume on column efficiency. Column: 0.25-mm ID × 98-mm length. Mobile phase: acetonitrile/water (70/30). Flow rate: 2.2 µl/min. Solutes and $k'$ values: □, benzene ($k' = 1.2$); △, biphenyl ($k' = 2.9$); ○, pyrene ($k' = 7.3$).

valve injector (JASCO) allows employment of 0.25-mm ID microcolumns with minimum band broadening.

### 2.3.5. Detector Cell Volume

In order to efficiently detect a narrow peak from a small-volume column, a detector with a sufficiently small volume cell must be used.

When complete flow mixing occurs in the detector cell, the detector variance $\sigma_d^2$ in terms of volume is

$$\sigma_d^2 = V_d^2 \tag{18}$$

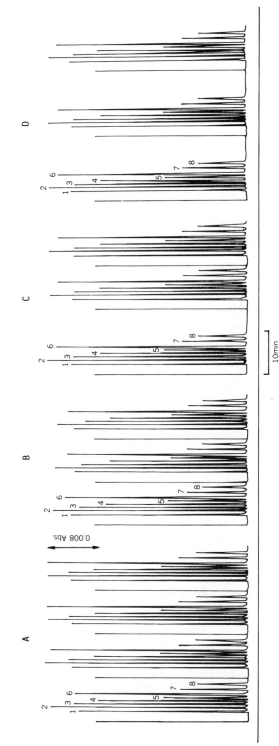

**FIGURE 2-10.** Reproducibility of the injector. Column: each one of four columns (A–D) has the same dimensions of 0.25 mm ID × 10 mm long fused-silica tube packed with SC-01 5 μm ODS. Mobile phase: acetonitrile/water (70/30). Flow rate: 3 μl/min. Detector: UVIDEC-100-III at 254 nm. Figure shows repeated injections (three or four times each) into four columns (A–D) with the same dimensions. Samples: 1, benzene; 2, naphthalene; 3, biphenyl; 4, fluorene; 5, phenanthrene; 6, anthracene; 7, fluoranthene; 8, pyrene. Sample volume: 0.02μl.

**TABLE 2-6**
Maximum Detector Cell Volumes for Various Peak
Volumes[a]

| Maximum cell volume ($\mu$l) | Peak volumes ($\mu$l) |
|---|---|
| 0.1 | 1 |
| 1 | 10 |
| 10 | 100 |

[a]Cell volumes give an 8% increase in the peak volume if complete flow
mixing occurs.

where $V_d$ is the detector cell volume. With plug flow, the contribution of the detector cell should be

$$\sigma_d^2 = V_d^2/12 \qquad (19)$$

The detector cell will not work as a perfect mixing chamber; there is only partial remixing of the solute and the contribution of an actual detector cell is usually intermediate between these two extremes.[13] The detector variance increases as the flow rate increases and approaches the upper limit ($V_d^2$). Practically, as long as the detector cell volume ($V_d$) is less than about one-tenth of the volume of the peak of interest ($V_p$), extracolumn broadening by the detector will be insignificant.[14] In this case, even if complete mixing should occur in the detector cell, the increase in the peak volume is less than 8%.

Table 2-6 gives the maximum cell volume for various peak volumes. Thus, the maximum cell volumes for various columns of different diameters but having the same length ($L = 250$ mm), efficiency ($N = 10,000$), and porosity ($\epsilon = 0.7$) will be, assuming that $k' = 0$ (unretained peak),

$d_c = 4.6$ mm; maximum cell volume: 10 $\mu$l
$d_c = 1.5$ mm; maximum cell volume: 1 $\mu$l
$d_c = 0.5$ mm; maximum cell volume: 0.1 $\mu$l

In the case of a UV/VIS photometric detector, once the maximum cell volume has been determined, the cell path length ($l_{dc}$) should be made as long as possible, in order to keep what is gained in the peak concentration by the reduction of the column volume. The maximum value of $l_{dc}$ retains maximum sensitivity, since absorbance is proportional to the path length. On the other hand, optical transmittance of the cell decreases rapidly with decreasing cell aperture width, resulting in poor linearity and signal-to-noise ratio. For a cylindrical cell, the cell path length ($l_{dc}$) is hyperbolically related to the square of the cell diameter $d_{dc}^2$ for a given cell volume $V_d$.

$$l_{dc} = \frac{4V_d}{\pi d_{dc}^2} \qquad (20)$$

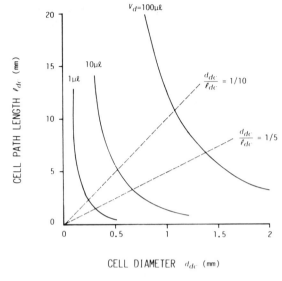

**FIGURE 2-11.** The path length related hyperbolically to the square of the cell diameter. Broken lines show the optical aperture ranging from 1/10 to 1/5. Outside this range, the utilizable light energy rapidly decreases or the path length becomes too short, resulting in poor linearity and/or signal-to-noise ratio.

This relationship is shown for various cell volumes in Fig. 2-11. Broken lines show the optical aperture ($d_{dc}/l_{dc}$) ranging from 1/10 to 1/5. The optics for LC detectors usually have numbers within this range: below one-tenth the utilizable light energy rapidly decreases, resulting in too much noise, while above one-fifth the path length becomes too short, resulting in too small signal. Thus, values of the cell diameter and path length in the domain closed by the broken lines (Fig. 2-11) will be practical.

### 2.3.6. Detector Time Constant

The detector time constant also causes peak broadening and it is a critical factor in high-speed analysis. Assuming that the time constant is added by a simple RC series circuit shown in Fig. 2-12, the detector time constant is treated as an exponential broadening factor applied to a Gaussian

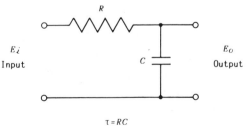

**FIGURE 2-12.** The RC low-pass filter reduces high frequency noise of a detector output and, at the same time, causes broadening of the time-based peak width.

**TABLE 2-7**
Maximum Time Constants for Peaks of Various Retention Times[a]

| Retention time (sec) | Maximum time constants ($\tau_{max}$) in sec for different theoretical plate number ($N$) values | | |
| --- | --- | --- | --- |
| | N = 5,000 | N = 10,000 | N = 20,000 |
| 6 | 0.013 | 0.009 | 0.006 |
| 60 | 0.13 | 0.09 | 0.06 |
| 180 | 0.38 | 0.26 | 0.19 |

[a]A 1% increase in the peak width is allowed.

input.[11,38,39] The time variance of the observed peak ($\sigma^2_{t(ob)}$) can be expressed as

$$\sigma^2_{t(ob)} = \sigma^2_t + \tau^2 \tag{21}$$

where $\sigma^2_t$ is the peak variance without the contribution of the time constant and $\tau$ is the $RC$ detector time constant. If 1% extra peak broadening is allowed, the maximum time constant ($\tau_{max}$) can be calculated (by inserting equation 21 into equation 11) to be about 0.14 times the original peak standard deviation ($\tau/\sigma_t = 0.14$) or about 4% of the peak width in terms of time. Table 2-7 lists maximum time constants for peaks having various retention times.

### 2.3.7. Connecting Tubing

As demonstrated so far, in microscale HPLC, even the smallest dead volume must be eliminated from the system. The connecting tubing must also be down-scaled in order to obtain the maximum efficiency from the microcolumn to be used.

The contribution of the Poiseuille flow dispersion to the peak variance ($\sigma^2_{pt}$) in long straight tubing can be derived from Taylor's work[25,26] and is proportional to the fourth power of the internal tube radius, the tube length, and the flow rate. The Taylor–Golay equation[27] for the plate height ($H$) of an open-tubular column is directly applicable for the calculation of the peak variance of the connecting tube.[11]

$$H = \frac{2D_M}{u} + \frac{r_o^2 u}{24D_M} \tag{22}$$

Where $D_M$ is the molecular diffusion coefficient in the mobile phase, $r_o$ is the internal radius of the tube, and $u$ is the linear velocity of the mobile-phase flow. The plate height of the tube ($H$) is $l_t/(\pi r_o^2 l_t/\sigma_{pt})^2$, so the peak variance can be expressed as

$$\sigma^2_{pt} = \pi^2 r_o^4 l_t H \tag{23}$$

Equation 23 can be rewritten by using equation 22 and substituting $F/\pi r_o^2$ for $u$.

$$\sigma_{pt}^2 = \frac{2\pi^3 D_M r_o^6 l_t}{F} + \frac{\pi r_o^4 l_t F}{24 D_M} \tag{24}$$

Molecular diffusion ($D_M$) in a liquid is very small, usually $1 \times 10^{-5}$ cm²/sec, and tube internal radius is also small, less than $2.5 \times 10^{-2}$ cm. The first term of equation 24 is therefore negligible. Thus, we obtain

$$\sigma_{pt}^2 = \frac{\pi r_o^4 l_t F}{24 D_M} \tag{25}$$

where the peak variance ($\sigma_{pt}^2$) is proportional to the fourth power of the internal tube radius ($r_o^4$), to the tube length ($l_t$), and to the flow rate ($F$).

In this way, the contribution of the connecting tube to the peak volume can be calculated. For easy comparison, equation 25 is again rewritten, by using $\sigma_{pt}^2 = (V_{pt}/4)^2$ and $S_t = \pi r_o^2$ (cross-sectional area), as

$$V_{pt}^2 = \frac{2 S_t^2 l_t F}{3 \pi D_M} \tag{26}$$

If the observed peak is allowed to have a volume greater by 3% than the column peak volume due to peak broadening in the connecting tube, the necessary cross-sectional area and length can be calculated. By inserting equation 26 into equation 11, we obtain

$$(1.03)^2 = 1 + \frac{2 S_t^2 l_t F}{3 \pi D_m V_p^2} \tag{27}$$

which can be rewritten as

$$S_t^2 l_t \cong 0.287 D_M V_p^2 / F \tag{28}$$

For $V_p = 1.0 \times 10^{-2}$ cm³ and $F = 1.67 \times 10^{-3}$ cm³/sec, which represent the typical volume of an early eluting peak, and the flow rate used in semi-micro-HPLC, assuming that $D_M = 1 \times 10^{-5}$ cm²/sec, we obtain

$$S_t^2 l_t = 1.72 \times 10^{-7} \tag{29}$$

for $V_p = 1.0 \times 10^{-3}$ cm³ and $F = 1.67 \times 10^{-4}$ cm³/sec, which are also the typical peak volume, and the flow rate in micro-HPLC, assuming that $D_M = 1 \times 10^{-5}$ cm²/sec, we obtain

$$S_t^2 l_t = 1.72 \times 10^{-8} \tag{30}$$

If a connecting tube with an internal diameter of $1 \times 10^{-3}$ cm is used, the maximum tube lengths should be

$l_{tmax} \simeq 28$ cm for a semi-microcolumn
$l_{tmax} \simeq 2.8$ cm for a microcolumn

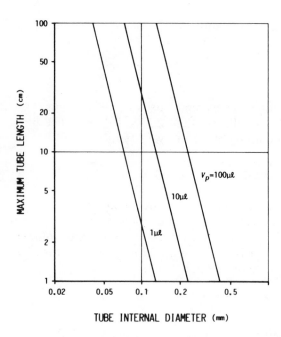

**FIGURE 2-13.** Maximum tube lengths and diameters allowed for 3% broadening for peaks having various volumes ($V_p$). The flow rates are assumed to be 1000, 100, and 10 $\mu$l/min for 100-, 10-, and 1-$\mu$l peak volumes, respectively.

If a connecting tube with an internal diameter of $5 \times 10^{-4}$cm is used, the maximum tube length, even for the microcolumn specified above, will be

$$l_{tmax} \simeq 45 \text{ cm}$$

It should be noted that if a conventional connecting tube of 0.25 mm ID is to be used in micro-HPLC, the maximum tube length would be $1/(2.5)^4$ times the above values, i.e., 0.7 cm for a semi-microcolumn and only 0.07 cm for a microcolumn. These values are obviously impractical.

Figure 2-13 shows the relationship between the maximum tube lengths and the tube diameters allowed for 3% increase in various peak volumes. As can be seen, without using tubing with an internal diameter of 0.1 mm or less, the connection between the injector, the column, and the detector becomes impossible in micro-HPLC. The use of 0.05 mm ID tubing permits the connection between even a microcolumn and its peripheral devices. Connecting the column directly to the detector cell and injector was previously proposed. [7, 33] However, for system versatility, there is no need, even in micro-HPLC, to hesitate to use connecting tubing as long as its internal diameter is less than 0.1 mm for a semi-microcolumn, and less than 0.05 mm for a microcolumn. Naturally, the tube length should be as short as possible and even the smallest dead volumes in the fittings must be eliminated. Table 2-8 lists maximum tube lengths for various peak volumes.

It must be emphasized that the calculations given above are valid only for long, straight tubing, and the contribution of curved tubing, which is common in a practical HPLC system, is appreciably less than that of straight tubing, especially in the high linear velocity range.[1,17,18,22-24,40] The reduction

**TABLE 2-8**
Maximum Tube Lengths for Various Peak Volumes[a]

| | | Tube ID (mm) | | |
| | | 0.25 | 0.1 | 0.05 |
| Peak volume (μl) | Flow rate (μl/min) | Maximum tube length (cm) | | |
|---|---|---|---|---|
| 1 | 10 | 0.07 | 2.8 | 45 |
| 10 | 100 | 0.7 | 28 | 450 |
| 100 | 1000 | 7 | 280 | 4500 |

[a]A 3% increase in the peak width is allowed.

of peak broadening is caused by the inertial flow that involves radial mixing. On the other hand, in the case of Poiseuille flow in straight tubing, the radial mixing is caused only by diffusion of the sample solute in the mobile phase. Figure 2-14 shows peak variances of a straight tube and a coiled tube for various flow rates.

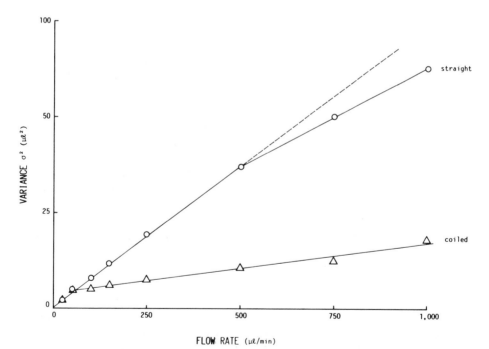

**FIGURE 2-14.** Peak variance contribution of a coiled tube is appreciably less than the contribution of a straight tube. Variances of the coiled tube (0.1-mm ID × 1000-mm length, coil diameter about 23 mm) and the straight tube having the same dimensions are plotted as a function of the flow rate. Injector: ML-425 with a 1-μl loop. Detector: UVIDEC-100-V with 1-μl cell. Sample: pyrene 32 ng/μl. The tubes were directly connected to the injector and detector with known system variance. The total variances of the tubes were calculated by measuring $\sigma$ at 0.607 of the peak height and then subtracting the system variance at each flow rate.

## 2.4. HPLC Components Compatible with Semi-Microcolumns and Microcolumns

HPLC components, pumps, injectors, and detectors that are compatible with microcolumns and semi-microcolumns are now commercially available from many manufacturers.[41,42]

The commercially available HPLC pumps can be categorized as two types: reciprocating pumps and syringe pumps. At present, reciprocating pumps are used most widely in HPLC. They are easy to use and provide an unlimited solvent supply to the column. Usually such pumps cover flow rate range of 10 to 1000 $\mu$l/min and 0.1 to 10.0 ml/min, thus permitting semi-micro, conventional, and semi-preparative applications and also fast HPLC analysis with a high linear mobile-phase velocity. Syringe pumps were used in the early stage of HPLC during the 1960s and the early 1970s. Old syringe pumps had a volume of 200–500 ml; modern syringe pumps have a volume of 0.5–50 ml and are used with relatively low flow rates. A syringe pump delivers a true pulse-free flow. However, due to the limited syringe capacity, it does not suitably cover the flow rate range of conventional columns. Usual flow rate range of a modern syringe pump is about 1 $\mu$l/min to 2 ml/min.

The commercial detectors that can be used in connection with semi-microcolumns and microcolumns are limited to only UV detectors. It is convenient to have a detector with a construction that allows an easy exchange of the detector cell. In this way, different cells, such as a 1-$\mu$l cell with a 5-mm path length, a conventional 8-$\mu$l cell with a 10-mm path length, and an 8-$\mu$l preparative cell with 1-mm path length, can be used. Similarly, the detector time constant can also be selected, e.g., from 1 to 0.05 sec.

Microvalve injectors are also available from several commercial sources. They have internal loop type configuration with the loop volume ranging from 0.02 to 1 $\mu$l. This type of injectors is not generally allowed to change the injection volume by partial filling of the loop using a microsyringe, therefore, several different loops must be prepared if the operator needs to change the injection volume.

For connecting tubing, a stainless steel tube of 0.1-mm ID by 1/32 or 1/16 inches can be used for semi-micro-HPLC. Ordinary tube fittings can be used for the connection of those tubes. For micro-HPLC, a fused-silica tube with 0.05 mm ID can be used.

If a chromatographer intends to assemble his own micro-HPLC or semi-micro-HPLC system by purchasing components, connections between these components must be carried out with great care in order to avoid excessive extracolumn peak broadening.

### References

1. Horváth, C.G.; Preiss, B.A.; Lipsky, S.R. *Anal. Chem.* **1967, 39,** 1422.
2. Horváth, C.G.; Lipsky, S.R. *Anal. Chem.* **1969, 41,** 1227.

3. Ishii, D.; Sakurai, K. "Tokyo Conference of Applied Spectrometry," Tokyo, Japan, October 1973, abstracts 1B05, p 73.
4. Ishii, D.; Asai, K.; Hibi, K.; Jonokuchi, T.; Nagaya, M. *J. Chromotogr.* **1977, 144,** 157.
5. Ishii, D.; Hibi, K.; Asai, K.; Jonokuchi, T. *J. Chromatogr.* **1978, 151,** 147.
6. Scott, R.P.W.; Kucera, P. *J. Chromatogr.* **1976, 125,** 251.
7. Scott, R.P.W.; Kucera, P. *J. Chromatogr.* **1979, 169,** 51.
8. Karasek, F.W. *Research/Development* **1977,** Jan p.42.
9. Knox, J.H. *J. Chromatogr. Sci.* **1977, 15,** 352.
10. Saito, M.; Waka, A.; Hibi, K.; Takahashi, M. *Industrial Research/Development,* **1983,** Apr, p.102.
11. Sternberg; J.C. *Advan. Chromatogr.* **1966, 2,** 205.
12. Scott, R.P.W.; Kucera, P. *J. Chromatogr. Sci.* **1971, 9,** 641.
13. Martin, M.; Eon, C.; Guiochon, G. *J. Chromatogr.* **1975, 108,** 229.
14. Kirkland, J.J.; Yan, W.W.; Stoklosa, H.J.; Dilks, C.H., Jr., *J. Chromatogr. Sci.* **1977, 15,** 303.
15. Coq, B.; Cretier, G.; Rocca, J.L. *J. Chromatogr.* **1979, 178,** 41.
16. DiCesare, J.L.; Dong, M.W.; Atwood, J.G. *J. Chromatogr.* **1981, 217,** 369.
17. Golay, M.J.E.; Atwood, J.G. *J. Chromatogr.* **1979, 186,** 353.
18. Atwood, J.G.; Golay, M.J.E. *J. Chromatogr.* **1981, 218,** 97.
19. Lauer, H.H.; Rozing, G.P. *Chromatographia* **1981, 14,** 641.
20. Wright, N.A.; Villalanti, D.C.; Burke, M.F. *Anal. Chem.* **1982, 54,** 1735.
21. Hupe, K.P.; Jonker, R.J.; Rozing, G. *J. Chromatogr.* **1984, 285,** 253.
22. Katz, E.D.; Scott, R.P.W. *J. Chromatogr.* **1983, 268,** 169.
23. Hofmann, K.; Halasz, I. *J. Chromatogr.* **1979, 173,** 211.
24. Hofmann, K.; Halasz, I. *J. Chromatogr.* **1980, 199,** 3.
25. Taylor, G. *Proc. Roy. Soc. London, Ser. A* **1953, 219,** 186.
26. Taylor, G. *Proc. Roy. Soc. London, Ser. A* **1954, 225,** 473.
27. Golay, M.J.E. In Goates, V.J.; Noebels, H.J.; Fagerson, I.S. Eds.; "Gas Chromatography 1957 (Lansing Symposium)", Academic Press: New York, 1958; p 1.
28. Colin, H.; Martin, M.; Guiochon, G. *J. Chromatogr* **1979, 185,** 79.
29. Harvey, M.C.; Stearns, S.D. *J. Chromatogr. Sci.* **1983, 21,** 473.
30. Coq, B.; Cretier, G.; Rocca, J.L. *J. Chromatogr. Sci.* **1981, 19,** 1.
31. Hirata, Y.; Novotny, M. *J. Chromatogr.* **1979, 186,** 521.
32. Tsuda, T.; Novotny, M. *Anal. Chem.* **1978, 50,** 632.
33. Yang, F.J. *J. Chromatogr.* **1982, 236,** 265.
34. McGuffin, V.L.; Novotny, M. *Anal. Chem.* **1983, 55,** 580.
35. Takeuchi, T.; Ishii, D. *J. High Resolut. Chromatogr./Chromatogr. Commun.* **1981, 4,** 469.
36. Ishii, D.; Konishi, H. U.S. Patent 4 346 610, 1982.
37. Takeuchi, T.; Ishii, D. *J. Chromatogr.* **1981, 213,** 25.
38. Schmauch, L.J. *Anal. Chem.* **1959, 31,** 225.
39. McWilliam, I.G.; Bolton, H.C. *Anal Chem.* **1969, 41,** 1755.
40. Tijssen, R. *Separ. Sci. Technol.* **1978, 13,** 681.
41. International Chromatography Guide, *J. Chromatogr. Sci.* **1986, 24,** 1G.
42. 1986 Buyers' Guide Edition, International Laboratory: Fairfield, Connecticut **1986.**

# 3

# Microscale Columns

## D. Ishii, T. Takeuchi, and A. Wada

### 3.1. Introduction

As discussed in Chapter 1, microcolumns can be divided into several types, according to the packing state and column dimensions. This chapter will cover the characteristics of microcolumns with 0.1–0.5 mm ID, open-tubular columns and packed microcapillary columns. The volume of these microcolumns is $\frac{1}{2000}$ to one-hundredth less than that of a typical conventional HPLC column with 4.6-mm ID and a 250-mm length. The necessary instrumentation, the preparation and characteristics of microcolumns, a pretreatment method, and the usefulness of exotic mobile phase in micro-HPLC will all be discussed in this chapter.

### 3.2. Necessary Instrumentation

### 3.2.1. Pumps

In order to achieve the same retention time independent of the column diameter, the flow rate must be adjusted so that the linear velocity is kept constant. Micro-HPLC therefore requires a pump that provides flow rates between 1 to 10 $\mu$l/min. As mentioned in Chapter 2, high-pressure micropumps are now commercially available. A microfeeder equipped with a gas-tight syringe may also work as a pump for micro-HPLC, and it can withstand an approximately 70 bar pressure. The flow rate of such a syringe pump can be changed by the dimensions of the syringe and flow rates of less than 1 $\mu$l/min can be generated with good precision.

*D. Ishii and T. Takeuchi* Department of Applied Chemistry, Faculty of Engineering, Nagoya University, Chikusa-ku, Nagoya 464, Japan. *A. Wada* Japan Spectroscopic Company, Ltd., Hachioji City, Tokyo 192, Japan.

### 3.2.2. Injectors

The injection volume should be carefully selected with consideration given to the column dimensions. When a microcolumn with a 0.25-mm ID and 100-mm length is employed, the injection volume should be reduced to approximately 0.02 $\mu$l, as discussed in Chapter 2. Such a small volume can be loaded with good reproducibility [e.g., by the ML-422 microvalve injector (JASCO)] and the injection volume can be altered by changing the dimension of the hole in the rotor disk. Several disks with different volume are provided for this injector, Rheodyne 7520* and EYELA 5001† (Tokyo, Japan) valve injectors are based on the same operating principle as the JASCO ML-422 injector. It should be noted that six different injection volumes (0.1, 0.2, 0.3, 0.4, 0.5, and 1 $\mu$l) can be selected with a single EYELA 5001 injector.

The connecting tubing between the column and the injector also plays as important role to obtain satisfactory results. Narrow-bore tubing with around 50- to 70-$\mu$m ID should be employed as the connecting tubing in order to reduce the extracolumn band broadening. The above-mentioned injectors are designed to be fitted with $\frac{1}{16}$ inch outer diameter (OD) tubing, which is not commercially available with such a narrow inner diameter. However, a $\frac{1}{16}$ inch OD stainless-steel tube, in which narrow-bore, fused-silica, or stainless steel tubing is cemented, works well as the connecting tubing between the column and the injector, although it is a somewhat delicate job to prepare it.

### 3.2.3. Detectors

Various types of detectors have been employed in HPLC. Among them the UV/visible, fluorescence electrochemical detectors, and refractometers are the most common. Possibilities and problems encountered when employing these detectors in micro-HPLC will be discussed in Chapters 4, 5, and 6. This chapter will only briefly mention the use of UV/visible detectors in micro-HPLC.

The flow cell used in the optical detector can be divided into two types according to its structure: the parallel-flow cell and the cross-flow cell, as shown in Fig. 3-1. The former is capable of producing higher sensitivity, but preparation of such a microflow cell is difficult. The latter uses a cylindrical quartz as the flow cell and the dead volume of the connection between the separation column and the flow cell can be easily minimized. The diameter of the cross-flow cell is nearly of the same dimension as that of the separation column. Although it is not as versatile, the exit of the separation column can also work as the flow cell when a fused-silica tube is employed as the column.[1]

---

*6815 So. Santa Rosa Ave., Cotati, CA 94928
†Muromachi 4-2, Nihonbashi, Chuo-ku, Tokyo, Japan

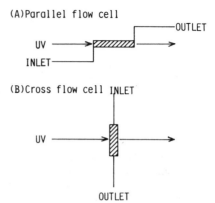

**FIGURE 3-1.** Structures of flow cells.

A multichannel, photodiode array UV/visible detector can also be applied in micro-HPLC by decreasing the number of optical parts, such as lenses and mirrors, and by reducing the distance between the light source and the flow cell and between the flow cell and the grating.[2]

### 3.2.4. Gradient Equipment

Solvent-gradient elution is effective in reducing the analysis time and improving the selectivity in HPLC. Its use has also been examined in micro-HPLC.[3,4] Although micropumps that allowing gradient elution to be carried out are commercially available, the reproducibility or accuracy of gradient elution at flow rates lower than several tens of $\mu$l/min is not satisfactory. A gradient method that uses a small-volume mixing chamber is effective in micro-HPLC. In this method, the gradient profile varies exponentially with time ($t$), depending on the flow rate ($F$) and the volume of the mixing chamber ($V_m$),[3] as indicated by

$$x = a - (a - x_o)\exp\left(\frac{-Ft}{V_m}\right) \qquad (1)$$

where $a$ is the concentration of the incoming solution and $x_o$ is the initial concentration of the solution in the mixing chamber. Figure 3-2 illustrates the block diagram of the gradient elution system using a mixing chamber. The system permits performance of gradient elution at low flow rates (less than 10 $\mu$l/min).

### 3.3. Precolumn Concentration

Increased mass sensitivity due to miniaturization of the columns is useful in trace analysis and is especially favorable in the analysis of precious samples. However, valve injection spoils most of the samples and valve injection is not preferred in the analysis of valuable samples, such as com-

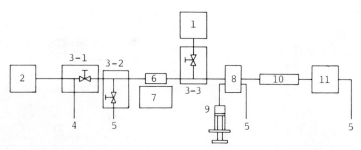

**FIGURE 3-2.** Block diagram of the gradient-elution system. 1, pump A (Model 590, Waters Associates, 34 Maple Street, Milford, MA 01757); 2, pump B (microMetric metering pump, LDC/Milton Roy, P.O. Box 10235, Riviera Beach, FL 33404); 3, three-way valves; 4, plug; 5, to drain; 6, mixing chamber; 7, magnetic stirrer; 8, microvalve injector; 9, gas-tight syringe; 10, column; 11, UV detector.

ponents present in blood serum. The precolumn concentration method is an effective solution to this problem. In this method, the adequate volume of sample solution is passed through the micro precolumn prior to the chromatographic run and all the concentrated solutes in the precolumn are then subjected to chromatographic separation.[5] Solutes of interest can be effectively concentrated in the precolumn by selecting the proper packing material and altering the property of the matrix solution. In addition, the precolumn concentration method overcomes the drawback of the low concentration sensitivity of micro-HPLC. When the concentration of a solute is 1 ppb, 1 ng of the solute is subjected to the separation if 1 ml of the sample solution is passed through the precolumn and recovery is perfect.

An online precolumn concentration system can be constructed by using the N6W switching valve.* Figure 3-3 shows an on-line precolumn concentration system consisting of two pumps, three switching valves, two precolumns, a separation column, and a detector. The system with two precolumns allows concentration of the sample solution in one column, while the other column is used for analysis. In order to minimize sample band broadening in the parts between the separation column and the precolumn, narrow-bore tubing should be employed as the connecting tubing to the N6W switching valve. Off-line precolumn concentration is also applicable in micro-HPLC, although it involves complicated procedures.

The dimension of the micro precolumn are 0.2–0.5-mm ID × 10–30-mm length, which is about one-tenth of the dimensions of the separation column. The particle diameter of the packing in the micro precolumn should be relatively large (15–30 $\mu$m), as such packing has a good permeability and allows for high-speed concentration. Stainless-steel tube fittings are favored because of the high-pressure operation. Precolumns made of PTFE tubing can be used at pressures lower than 70 bar.

*Valco, P.O. Box 55603, Houston, TX 77255.

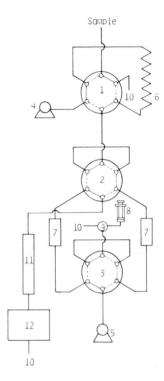

**FIGURE 3-3.** Diagram of a precolumn concentration system. 1, switching valve no. 1 (Rheodyne 7000); 2, switching valve no. 2 (Valco N6W); 3, switching valve no. 3 (Rheodyne 7000); 4, pump (Familic-300S); 5, pump (LKB 2150); 6, sample loop; 7, concentration columns; 8, gas-tight syringe for the measurement of the sample volume; 9, three-way stopcock; 10, drain; 11, separation column; 12, UV detector.

## 3.4. Characteristics of Microcolumns

The column preparation techniques that are utilized in conventional HPLC are also applicable to microcolumns if the tube fitting withstands high pressure. Hydrophobic columns, such as octadecyl silica (ODS), can be prepared by the viscosity method or the slurry-packing method with a nonionic detergent.

### 3.4.1. Column Material

Various types of column tube materials such as PTFE, stainless-steel, glass, fused-silica tubing, or glass-lined stainless-steel tubing have been utilized in micro-HPLC. Among these the fused-silica and glass columns give higher column efficiencies, due to their smooth and inert surface. Fused-silica tubing is flexible, which is convenient for handling long columns. Polyimide ferrules that permit high-pressure operation are commercially available for fused-silica tubing. When fused-silica tubing is glued in a stainless-steel tubing, tube fitting with stainless-steel ferrules and unions is possible.[6] Glass-lined stainless-steel tubing with different inner diameters is also commercially available. Glass-lined stainless-steel tubing with a 1/16 inch OD is capable of connecting the column with commercially available unions,

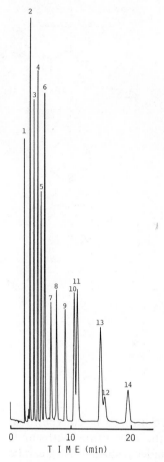

**FIGURE 3-4.** Separation of PAHs on a fused-silica column. Column: ODS SC-01, 0.25-mm ID × 100-mm length. Mobile phase: acetonitrile/water (7/3). Flow rate: 3.3 μl/min. UV detector used at 254 nm. Temperature: 24°C. Peaks: 1, benzene; 2, naphthalene; 3, biphenyl; 4, fluorene; 5, phenanthrene; 6, anthracene; 7, fluoranthene; 8, pyrene; 9, *p*-terphenyl; 10, chrysene; 11, 9-phenylanthracene; 12, perylene; 13, 1,3,5-triphenylbenzene; 14, benzo(a)pyrene.

facilitating the preparation and operation of the column under high pressures.

Figure 3-4 shows the separation of polynuclear aromatic hydrocarbons (PAHs) on a 0.25-mm ID × 100-mm long ODS column.[7] It has been found that microcolumns made of fused-silica tubing can achieve column efficiencies comparable to those of conventional columns.

### 3.4.2. High-Speed Separations

High-speed separations have become increasingly important in HPLC.[8-12] and offer several advantages over the conventional methods, such as decreased analysis time and cost and the possibility of analyzing unstable samples. Such separations can generally be achieved using a short column packed with fine particles. One of the advantages of micro-HPLC is the possibility of achieving high-speed separation at low flow rates. The flow rates required for high-speed separations in micro-HPLC are 10–50 μl/min, and most of the recently available HPLC pumps can achieve such flow rates.

Tube fittings and connecting tubing between each component should be carefully designed for high-speed separation in order to minimize extra-column band broadening, because a short column with a 3–10-cm length is generally employed.[6]

When 1600 theoretical plates are achieved on a 5 cm column at the linear velocity of 1 cm/sec, the elution time and the peak width of an unretained solute should be 5 sec and 0.5 sec, respectively. This is the reason why both a detector and a recorder with a fast response must be used. The time constant of UV/visible detectors is commonly 1 sec, which causes band broadening for narrow-band peaks. The time constant should be less than 50 msec in the case of high-speed separations. UV/visible detectors that can easily select different time constants are available.

Figure 3-5 demonstrates separations of PAHs on a 0.34 mm ID × 50 mm long column obtained with two different time constants (0.05 sec and 1 sec). It is evident that the effect of the response time on the separation is significant.

Figure 3-6 shows the relationship between the theoretical plate height (HETP) and the linear mobile phase velocity (*u*) for 3 and 5 μm ODS columns. The 3-μm columns have the ability to achieve a separation faster than the 5-μm columns. The high-speed separation of PAHs is demonstrated in Fig. 3-7, in which the separation of 14 PAHs was achieved in 2 min.

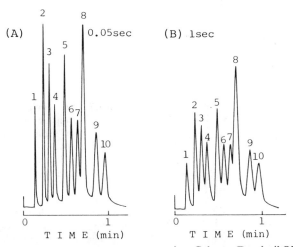

**FIGURE 3-5.** Effect of the time constant on a separation. Column: Develosil ODS-3, 0.34-mm ID × 50-mm length. Mobile phase: acetonitrile/water (65/35). Flow rate: 30 μl/min. UV detector used at 254 nm. Peaks: 1, phenol; 2, benzene; 3, toluene; 4, naphthalene; 5, biphenyl; 6, fluorene; 7, phenanthrene; 8, anthracene; 9, fluoranthene; 10, pyrene.

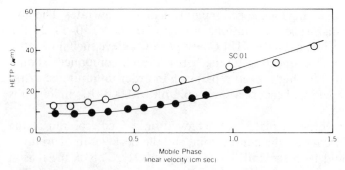

**FIGURE 3-6.** Effect of particle diameter on the relationship between HETP and linear mobile phase velocity. Column: ○, ODS SC-1; ●, Develosil ODS-3; 0.34-mm ID × 100-mm length. Mobile phase: acetonitrile/water (7/3). Sample: pyrene. Time constant: 0.04 sec.

### 3.4.3. High-Resolution Columns

A column that produces a large theoretical plate number for the separation of complex mixtures of solutes with quite similar retention times should be employed. For two solutes, the resolution ($R_s$) can be expressed by the following equation:

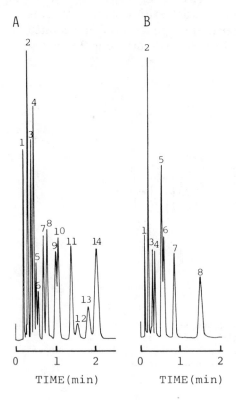

**FIGURE 3-7.** High-speed separation of PAHs. Column: Develosil ODS-3, 0.34-mm ID × 50-mm length. Mobile phases: acetonitrile/water (A) (7/3) and (B) (8/2). Inlet pressure: 150 kg/cm². Flow rates: (A) 30 μl/min; (B) 40 μl/min. Time constant: 0.04 sec. Peaks: (A) 1, benzene; 2, naphthalene; 3, biphenyl; 4, fluorene; 5, phenanthrene; 6, anthracene; 7, fluoranthene; 8, pyrene; 9, p-terphenyl; 10, 9-phenylanthracene; 11, chrysene; 12, perylene; 13, benzo(a)pyrene; 14, 1,3,5-triphenylbenzene. (B) 1, benzene; 2, fluorene; 3, pyrene; 4, chrysene; 5, 1,3,5-triphenylbenzene; 6, benzo(a)pyrene; 7, benzo(g,h,i)perylene; 8, coronene. Wavelengths: (A) 254 nm; (B) 280 mm.

$$R_s = \frac{\alpha - 1}{4} \sqrt{N_{eff}} \qquad (2)$$

Here $\alpha$ is the separation factor ($\alpha = k'_2/k'_1$) and $N_{eff}$ is the effective theoretical plate number. It can be expected from equation 2, that two solutes with an $\alpha$ equaling 1.013 can be separated by a column producing $10^5$ effective theoretical plates and those with an $\alpha$ equaling 1.004 can be separated by a column producing $10^6$ effective theoretical plates.

Micro-HPLC facilitates the use of long columns, resulting in high theoretical plate numbers that are proportional to the column length, due to the decreased multipath diffusion in the column and the effective transfer of the heat generated in the column. Fused-silica tubing is convenient for long columns because it is easy to handle due to its flexibility.

Figure 3-8 shows the relationship between the theoretical plate number and the column length in the size-exclusion chromatography. Sixty thousand (60,000) theoretical plates per meter column length are achieved when 5 $\mu$m packing materials are used.[13] Large theoretical plate numbers are also obtained in reversed-phase chromatography, using 1–2-m long columns.[1,14,15] It takes a great deal of time to carry out a separation with a long column in reversed-phase chromatography; size-exclusion chromatography requires considerably less time.

Figure 3-9 demonstrates the separation of the mixture of polychlorinated biphenyls on a 0.34-mm ID × 100-cm length ODS column in the reversed-phase mode. About 70,000 theoretical plates were achieved, which is still not enough to completely resolve such a mixture. Figure 3-10 shows the separation of phthalates on a 0.35-mm ID × 143-cm long column used in the size-exclusion mode. The latter mode has been successfully applied

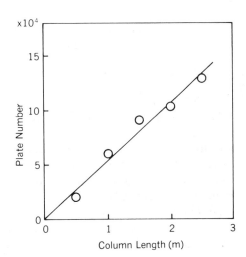

**FIGURE 3-8.** Relationship between column length and the theoretical plate number. Column: TSK-GEL G 3000H. Mobile phase: THF. Flow rate: 1.04 $\mu$l/min. Sample: benzene.

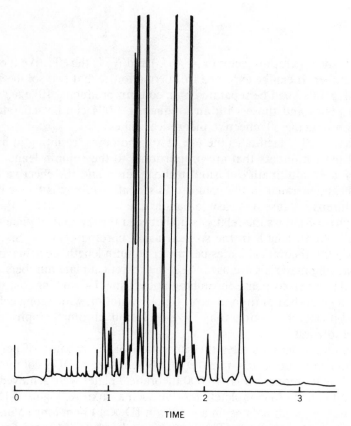

TIME

**FIGURE 3-9.** Separation of a mixture of PCBs on a 0.34-mm ID × 1-m long packed fused-silica microcolumn. Pump: FAMILIC-300. Column: SC-01. Mobile phase: acetonitrile/water (85/15). Inlet pressure: 80 kg/cm$^2$. Flow rate: 2.8 $\mu$l/min. Sample: 0.29 $\mu$l of a 1% PCB-48 solution (chlorine content 48%). UV detector used at 254 nm.

to the separation of oligomers of epoxy resins and phenol-formaldehyde resins, which will be discussed in Chapter 7.

### 3.4.4. Temperature Programming

Due to their small heat capacity, microcolumns facilitate the use of temperature programming.[15,16] Temperature programming does not require a complicated system, as compared with a solvent-gradient system. It only requires an oven in which the column temperature is programmable. The usual ovens used in gas chromatography can be utilized for this purpose. Temperature programming can replace a part of the separations of thermally stable solutes, which until now have been carried out by solvent-gradient elution.

Figure 3-11 shows the separation of *p*-nitrobenzyl esters of fatty acids obtained under isothermal and temperature-programming conditions.[16] It

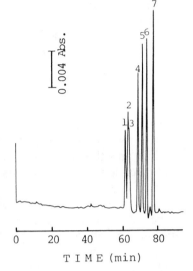

**FIGURE 3-10.** Separation of phthalates on a micro-column. Column: 0.35-mm ID × 142.6-cm length, packed with 1000H (5μm). Flow rate: 1.04 μl/min. Peaks: 1, dimethyl phthalate; 2, diethyl phthalate; 3, di-*n*-propyl phthalate; 4, di-*n*-butyl phthalate; 7, di-heptyl phthalate; 8, di-2-ethylhexyl phthalate; 9, dinonyl phthalate. Each phthalate is 0.2%

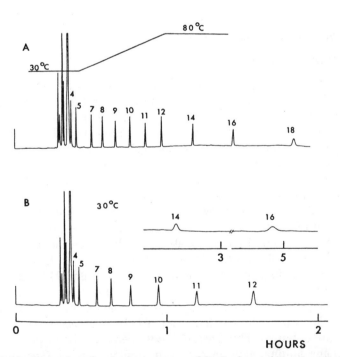

**FIGURE 3-11.** Separation of *p*-nitrobenzyl esters of $C_4$–$C_{18}$ fatty acids under (A) temperature gradient and (B) isothermal elution. Column: 0.2-mm ID × 1-m length, packed with 3 μm packing. Mobile phase: acetonitrile/water (90/10). Pressure: 300 kg/cm². UV detector used at 254 nm. The numbers above the peaks identify the carbon number of the fatty acids.

can be seen that the latter significantly reduces analysis time while it increases sensitivity.

### 3.4.5. Use of the Exotic Mobile Phase

Micro-HPLC facilitates the use of toxic, expensive, flammable, or exotic mobile phase. The use of low alkanes,[17] carbon dioxide,[18] and deuterated solvents[19,20] in micro-HPLC has been examined. Deuterated solvents can be used in coupling HPLC with infrared (IR) or nuclear magnetic resonance (NMR) spectrometry. Low alkanes or carbon dioxide have low viscosities and they therefore enable high-speed separations. In order to deal with such low boiling solvents as the mobile phase, the whole system must be modified to withstand pressures higher than their vapor pressures. When these low-boiling mobile phases are employed, the solutes are subjected to either LC or supercritical fluid chromatographic (SFC) separations, depending on the column temperature. The critical temperature (approximately 31°C) of carbon dioxide is close to room temperature, which allows both LC and SFC to be carried out using the same apparatus. Separations and characteristics obtained with these low-boiling solvents will be discussed in Chapter 7.

### 3.5. Open-Tubular (Capillary) Columns

### 3.5.1. Introduction

Open-tubular columns were originally proposed by Golay[21–23] for use in GC and have been widely utilized. In GC, open-tubular columns have a good permeability and consequently produce higher theoretical plate numbers per unit time and unit pressure drop across the column than packed columns. Open-tubular columns producing higher theoretical plate numbers can resolve complex mixtures, which are difficult to separate on conventional packed columns. Due to their high resolution, the usefulness of open-tubular columns in LC has been investigated.

Open-tubular columns, with the same dimension as in GC, were employed in LC, but they gave poor column efficiencies, due to the low diffusion speed in a liquid mobile phase.[24–28] It has been theoretically proven that narrow-bore columns should be employed in LC and Ishii's group[26] first showed the feasibility of open-tubular LC using 60-$\mu$m ID columns.

The diffusion of a solute passing through a straight tube as a laminar flow has been examined in detail by Taylor.[27] The magnitude of the diffusion of a solute in laminar flow is too large in LC, which necessitates the use of narrow-bore columns. Thus, research has been directed toward the modification of the laminar flow in the open-tubular column. Operation in the turbulent flow region,[28] flow segmentation with air plugs,[29,30] the use of helically-coiled columns[31] or deformed or wavy tubes,[32] and the use of electroosmosis flow[33] have been investigated. Band spreading of unretained solutes could be reduced by these approaches, while band spreading of retained solutes was found to be greater than calculated. Most recent research on open-tubular LC has dealt with narrow-bore columns.[34-38]

The basic equation for open-tubular LC was derived by Golay.[23]

$$H = \frac{2D_M}{u} + \frac{(11k'^2 + 6k' + 1)d_c^2}{96(1 + k')^2 D_M} u + \frac{2k'd_f^2}{3(1 + k')^2 D_s} u \quad (3)$$

$$= \frac{2D_M}{u} + (C_M + C_s)u \quad (4)$$

Here $H$ is the plate height, $u$ is the linear velocity of the mobile phase, $k'$ is the capacity factor, $d_c$ is the column diameter, $d_f$ is the thickness of the stationary phase film and $D_M$ and $D_s$ are the diffusion coefficients of a solute in the mobile and stationary phases, respectively. The first term on the right hand side of equations 3 and 4 is due to longitudinal molecular diffusion, while the second and third terms are due to the resistance to mass transfer in the mobile and stationary phases, respectively. The diffusion coefficient of a solute in a liquid is much smaller than that in a gas, resulting in large contribution of the second term ($C_M$) in LC.

Knox and Gilbert[39] introduced reduced parameters into open-tubular LC.

$$h = \frac{2}{\nu} + \frac{11k'^2 + 6k' + 1}{96(1 + k')^2} \nu + \frac{2k'}{3(1 + k')^2}\left(\frac{d_f}{d_c}\right)^2\left(\frac{D_M}{D_s}\right) \quad (5)$$

Here $h$ is the reduced plate height ($h = H/d_c$) and $\nu$ is the reduced linear velocity ($\nu = u\,d_c/D_M$).

Decreasing the contribution of the second term ($C_M$) to the plate height plays an important role in achieving good results with open-tubular columns. Figure 3-12 illustrates the dependence of $C_M$ on $k'$ as well as on the column diameter. It can be seen that $C_M$ increases with increasing capacity factor and column diameter. In order to reduce the contribution of the resistance to mass transfer in the mobile phase, narrow-bore open-tubular columns should be operated under conditions achieving large diffusivity, i.e., operated at higher column temperatures and with low-viscosity mobile phases.

Bristow and Knox[40] defined the separation impedance ($E$), which permits the comparison of the performance of different types of columns operated under different conditions (See Section 1.3.).

$$E = \frac{t_o}{N} + \frac{\Delta P}{N} \times \frac{1}{\eta} \quad (6)$$

$$= h^2\phi \quad (7)$$

Here $t_o$ is the elution time of an unretained solute, $N$ is the number of theoretical plates, $\Delta P$ is the pressure drop along the column, $\eta$ is the viscosity of the mobile phase, $h$ is the reduced plate height, and $\phi$ is the column resis-

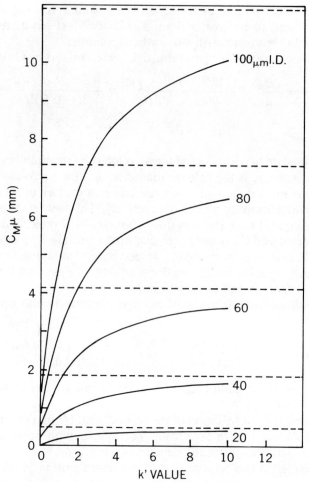

**FIGURE 3-12.**  Dependence of $C_M u$ on k' and the column diameter. $u$, 1 cm/sec; $D_M$, $1 \times 10^{-5}$ cm$^2$/sec. Each dotted line shows a convergent value of $C_M u$ as k' goes to infinity.

tance parameter. Open-tubular columns have a smaller value of $\phi$ compared with both densely- and loosely-packed columns. This means that open-tubular LC can achieve higher column efficiencies than HPLC when packed columns are used. Knox[41] theoretically calculated the column performance of the three types of columns and predicted that open-tubular columns, with plate numbers in excess of several hundred thousands, would provide a real competition for packed columns.

### 3.5.2. Instrumentation

Chromatographs for open-tubular LC are not available commercially. These columns have a quite small inner volume, e.g., a 50-$\mu$m ID $\times$ 5-m long column has a volume of about $10\mu$l. When such capillary columns are used as the separation columns, the extracolumn effect becomes serious. Flow rates lower than 1 $\mu$l/min are required in most cases in open-tubular LC. There is no LC pump that can supply the mobile phase at such a low flow rate. Both injection and detection volumes should be quite small, sometimes less than 10 nl. Thus, the apparatus must be designed by each laboratory.[26,35,36,38]

### 3.5.3. Wall-Coated Open-Tubular Columns

Both wall-coated and support-coated open-tubular columns have been commonly employed in GC. Although wall-coated columns have problems (the coated stationary phase is liable to dissolve in the liquid mobile phase) narrow-bore wall-coated columns are relatively easy to prepare and they are convenient to use in investigating the parameters that affect column efficiency. The dependence of column efficiency on the viscosity of the mobile and stationary phases, the column diameter, the thickness of the stationary phase film, the capacity factor, and the linear velocity can be predicted from equation 3. There are also other parameters that affect column efficiency, but their influence cannot be predicted from the basic equation. They include the pretreatment of the column, the injection volume, and the sample capacity.

Glass capillary tubing with small diameters can be prepared by using the commercially available glass drawing machines. The capillary tubing is then pretreated with an acid or alkaline solution and coated with the liquid phase. The dynamic coating method is more convenient for coating the liquid phase onto the narrow-bore capillary tubing than is the static coating method. Since the thickness of the stationary phase film in the dynamic method is affected by the coating speed, it is important to keep the speed constant in order to obtain good column efficiencies. The coating speed can be kept constant by attaching a buffer capillary tubing to the actual capillary column. The diameter of the buffer capillary should be the same as that of the actual column. It has been reported that pretreatment with an alkaline solution is effective to obtain high-efficiency columns for polar liquid phases.[42]

Figure 3-13 shows the separation of xylenols and *m*-cresol on a 33-$\mu$m ID $\times$ 8.13-m long column coated with $\beta$, $\beta'$-oxydipropionitrile. The mobile phase should be saturated with BOP, otherwise the stability of the column is rapidly reduced. The injection volume should be as small as possible when the solvent strength is higher than that of the mobile phase. The permitted injection volume increases when the solvent strength of the sample solution is the same or weaker than that of the mobile phase.

A large number of theoretical plates can be produced when the mobile phase velocity is decreased. Figure 3-14 shows the separation of aromatic amines on a 37 $\mu$m ID $\times$ 20 m long column coated with $\beta$, $\beta'$-oxydipropionitrile.[43] Seven hundred forty thousand (740,000) theoretical plates were

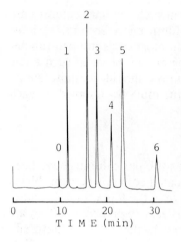

**FIGURE 3-13.** Separation of xylenol isomers and *m*-cresol on an open-tubular column. Column: 33-μm ID × 8.13-m length, coated with β,β'-oxydipropionitrile (BOP). Mobile phase: *n*-hexane saturated with BOP. Flow rate: 0.80 μl/min. Sample volume: 0.024 μl. UV detector utilized at 280 nm. Peaks: 0, solvent; 1, 2,6-xylenol; 2, 2,4-xylenol; 3, 2,3-xylenol; 4, 3,5-xylenol; 5, 3,4-xylenol; 6 = *m*-cresol.

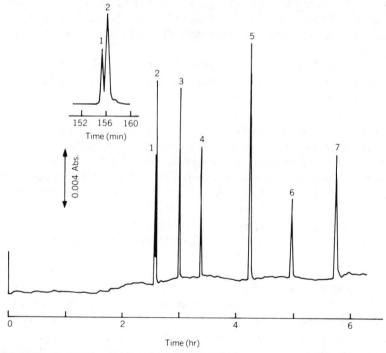

**FIGURE 3-14.** Separation of aromatic amines on an open-tubular column. Column: 37-μm ID × 20-m length coated with β,β'-oxydipropionitrile. Flow rate: 0.14 μl/min. UV detection used at 230 nm. Peak identification (k', N): 1, isooctane (0,740,000); 2, *N,N*-diethylaniline (0.006,610,000); 3, *N*-phenyl-α-naphthylamine (0.17,450,000); 4, *N*-phenyl-β-naphthylamine (0.31,360,000); 5, aniline (0.64,320,000); 6, α-naphthylamine (0.92,250,000); 7, β-naphthylamine (1.22,200,000).

produced for the unretained solute in 155 min and 200,000 theoretical plates were produced for the peak of $k' = 1.22$. The peak at $k' = 0.006$ (peak 2) is separated from the solvent peak ($k' = 0$).

It can be predicted from equation 6 that higher column efficiency can be achieved by using a mobile phase with low viscosity. Lower alkanes, such as propane and butane, have a lower viscosity than hexane. When an unretained solute passes through an open tube at a relatively high linear velocity, the first and the third term of equation 3 can be neglected and the square of the observed peak volumn ($V_{p(obs)2}$) can be expressed as

$$V_{p(obs)2} = \frac{\pi^2 d_c^6 L}{96 D_M} u \qquad (8)$$

where $L$ is the tube length. This equation is convenient to calculate $D_M$. Figure 3-15 shows the dependence of the square of the observed peak volume of an unretained solute on the linear velocity of the mobile phase.[44] The diffusion coefficient of the solute in various alkanes can be calculated from the results given in Figure 3-15. The diffusion coefficients of benzene in butane and propane are approximately $7 \times 10^{-5}$ and $1 \times 10^{-4}$ cm$^2$/sec, respectively, and they are 1.6 times and 2.3 times the magnitude of the value measured in $n$-hexane.

Figure 3-16 demonstrates separations of xylenol isomers on a short open-tubular column using $n$-butane and propane as the mobile phase.[44]

Figure 3-17 shows the rapid separation of xylenol isomers on a short

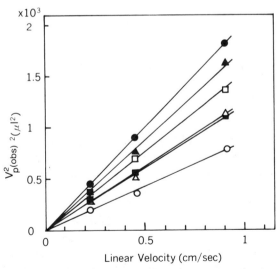

**FIGURE 3-15.** Dependence of the square of the observed peak volume of an unretained solute on the linear velocity. Column: 0.35-mm ID × 3.8-m length. Sample: benzene. Temperature: 18°C. Mobile phase: ●, $n$-hexane; ▲, neopentane; □, $n$-pentane; ■, $n$-butane; △, isobutane; ○, propane.

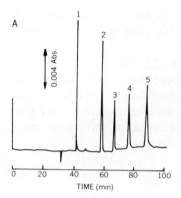

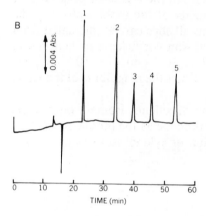

**FIGURE 3-16.** Separations of xylenol isomers on an open-tubular column, using low carbon-number alkanes as the mobile phase. Column: 37-$\mu$ m ID × 19.2-m length, coated with $\beta,\beta'$-oxydipropioni- trile. UV detector used at 280 nm. Mobile phase: (A), *n*-butane; (B), propane. Flow rates: (A), 0.69 $\mu$l/min; (B), 1.4 $\mu$l/min. Peaks: 1, 2,6-xylenol; 2, 2,5-xylenol; 3, 2,3-xylenol; 4, 3,5-xylenol; 5, 3,4- xylenol.

open-tubular column using propane as the mobile phase, with a linear veloc- ity of 11 cm/sec.

### 3.5.4. Solid-Phase Columns

It is well known in the field of GC with open-tubular columns that alka- line solutions have been employed to wash or etch the glass surface. For example, Nota et al.[25] treated a soft glass capillary with a 2.5 N sodium hydroxide solution for 2–8 hr at 100°C and then applied the prepared col- umn to the separation of dansyl-amino acids. Preparation conditions of such columns were examined in detail by Ishii et al.[45]

Treatment with an alkaline solution generates a silica gel layer on a glass surface and substantially increases the surface area of a capillary tube. Figure 3-18 shows scanning electron microphotos of a glass surface treated with an alkaline solution.[45] The structure, density, and hydrodynamic strength of these aggregates are dependent on the time and temperature of treatment and on the concentration and type of alkaline solution. Liquid

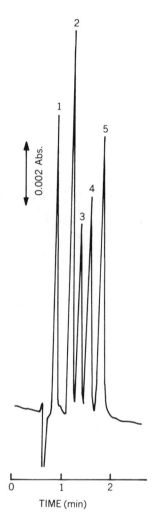

**FIGURE 3-17.** Rapid separation of xylenol isomers on a short open-tubular column using propane as the mobile phase. Column: 33-$\mu$m ID $\times$ 3.4-m length, coated with $\beta,\beta'$-oxydipropionitrile. Mobile phase: propane. Flow rate: 5.6 $\mu$l/min. Peaks and detector wavelength as in Figure 3-16.

stationary phases can be chemically bonded onto these aggregates; the characteristics of such chemically bonded columns will be described in Section 3.5.5.

Figure 3-19 demonstrates the separation of phenols on a solid phase column.[46] The column was treated with 1 N sodium hydroxide for 2 days at 54°C. With solid-phase columns, a wider variety of mobile phases can be utilized than with wall-coated columns.

### 3.5.5. Chemically Bonded Columns

There have been some papers published dealing with the preparation of chemically bonded stationary phases for open-tubular LC. Octadecyl sil-

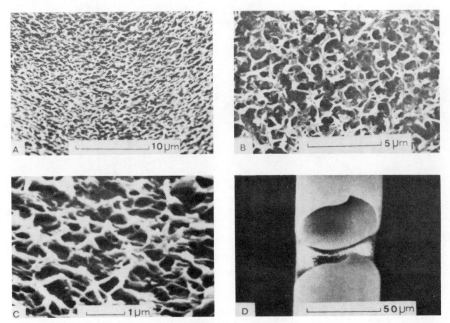

**FIGURE 3-18.** Scanning electron microphotos of a glass surface treated with an alkaline solution. Soda-lime glass treated with 1N sodium hydroxide solution for 2 days at room temperature (A, B, and C) or at 64°C (D). (A) magnification, ×3500; angle 60°. (B) Magnification, ×7500; angle 0°. (C) Magnification, ×20,000; angle 60°. (D) Magnification, ×750; angle 30°.

ane stationary phases were chemically bonded on the surface of fused-silica,[47] soft glass,[34] and borosilicate glass.[38] It was reported that the ODS columns prepared in this way were stable. Ishii and Takeuchi[48] reviewed the preparation procedures and the properties of nonextractable stationary phases for open-tubular LC, including chemically-bonded ODS and ion-exchange columns.

Due to the stability of the stationary phase, chemically bonded open-

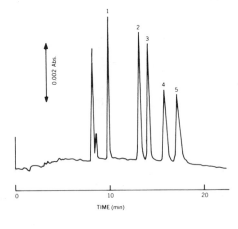

**FIGURE 3-19.** Separation of phenols on an open-tubular column with a silica gel layer. Column: 43-$\mu$m ID × 5.8-m length, treated with 1N sodium hydroxide solution for 2 days at 54°C. Mobile phase: *n*-hexane containing 0.2% ethyl acetate and 0.8% methanol. Flow rate: 1.1 $\mu$l/min. UV detector used at 280 nm. Peaks: 1, 2,6-xylenol; 2, 2,4-xylenol; 3, *O*-cresol; 4, 3,5-xylenol; 5, *m*-cresol.

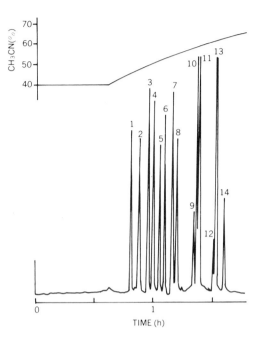

**FIGURE 3-20.** Solvent gradient-elution separation of PAHs on a bonded-phase open-tubular column. Column: 31-μm ID × 22-m length, with bonded ODS phase: Mobile phase acetonitrile/water, as indicated. Flow rate: 0.52 μl/min. Column temperature: 36°C. UV detector used at 250 nm. Peaks: 1, benzene; 2, naphthalene; 3, biphenyl; 4, fluorene; 5, phenanthrene; 6, anthracene; 7, fluoranthene; 8, pyrene; 9, *p*-terphenyl; 10, chrysene; 11, 9-phenylanthracene; 12, perylene; 13, benzo(a)pyrene; 14, 1,3,5,-triphenylbenzene.

tubular columns permit solvent-gradient elution to be carried out using packed columns, just as in HPLC. Such open-tubular columns also facilitate temperature programming.

Figures 3-20 and 3-21 show the separation of PAHs by solvent-gradient elution and temperature programming, respectively.[49]

Separation of nucleosides on a cation-exchange column is demonstrated in Figure 3-22, in which the difference in selectivity between aromatic and aliphatic cation-exchange columns is seen.[50]

### 3.5.6. Cross-Linked Columns

Polystyrene or polysiloxane stationary phases become nonextractable by cross linking. Cross-linked columns are applicable in open-tubular LC. Preparation and characteristics of cross-linked polystyrene[51] and polysiloxane[52] columns have been reported. The latter columns have recently been successfully employed due to their stability against heat and solvent extraction.

Figure 3-23 shows the solvent-gradient separation of phthalates on a cross-linked SE-54 column.[52]

### 3.5.7. Dynamically Modified Columns

By passing a mobile phase containing cationic or nonionic detergents through silica gel columns, you can dynamically modify the detergents onto the silica gel in situ and they can work as a hydrophobic stationary phase. Silica columns dynamically modified with long-chain quaternary ammo-

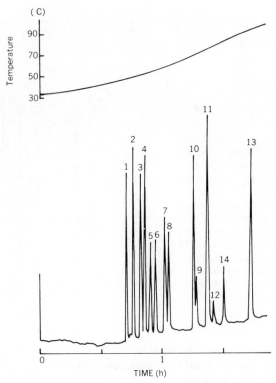

**FIGURE 3-21.** Temperature programming separation of PAHs on a bonded-phase open-tubular column. Column: as in Fig. 3-20. Mobile phase: acetonitrile/water (4/6). Inlet pressure: 40 kg/cm². UV detector used at 254 nm. Peaks: as in Fig. 3-20.

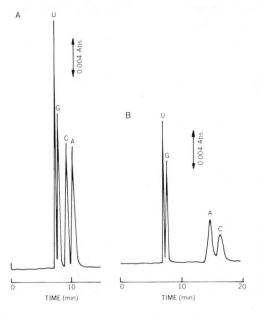

**FIGURE 3-22.** Separation of nucleosides on open-tubular columns with bonded cation-exchange groups. Columns: (A) 44-$\mu$m ID $\times$ 5.2-m length, $-C_6H_4SO_3H$; (B) 52-$\mu$m ID $\times$ 5.3-m length, $-C_2H_4SO_3H$. Mobile phase: ammonium formate, (A) $2 \times 10^{-3}$M, pH = 2.2; (B) $1 \times 10^{-3}$M, pH = 3.4. Flow rate: (A) 1.1 $\mu$l/min; (B) 1.7 $\mu$l/min. UV detector used at 260 nm. Peaks: U, uridine (12 ng); G, guanosine (12 ng); C, cytidine (13 ng); A, adenosine (11 ng).

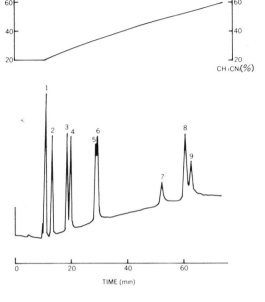

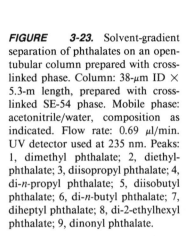

**FIGURE 3-23.** Solvent-gradient separation of phthalates on an open-tubular column prepared with cross-linked phase. Column: 38-$\mu$m ID $\times$ 5.3-m length, prepared with cross-linked SE-54 phase. Mobile phase: acetonitrile/water, composition as indicated. Flow rate: 0.69 $\mu$l/min. UV detector used at 235 nm. Peaks: 1, dimethyl phthalate; 2, diethyl-phthalate; 3, diisopropyl phthalate; 4, di-$n$-propyl phthalate; 5, diisobutyl phthalate; 6, di-$n$-butyl phthalate; 7, diheptyl phthalate; 8, di-2-ethylhexyl phthalate; 9, dinonyl phthalate.

nium ions work like chemically bonded ODS columns. The advantage of dynamically modified columns over chemically bonded columns is a slight variation in selectivity for different brands of column materials. The use of dynamically modified columns has also been investigated in open-tubular LC.[53]

Soda-lime glass capillaries are treated with a 1 N sodium hydroxide solution and passed through a mobile phase containing a hydrophobic layer-forming detergent, such as alkyltrimethylammonium bromide, until solute retention becomes constant. As the linked detergents cannot easily be released from the silica surface, it is possible to carry out gradient elution. Figure 3-24 shows the gradient separation of PAHs on a dynamically mod-

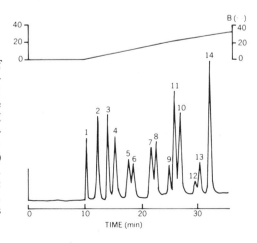

**FIGURE 3-24.** Gradient separation of PAHs on a dynamically-modified open-tubular column. Column: 35-$\mu$m ID $\times$ 7.8-m length, treated at 45°C. Mobile phase: (A) acetonitrile/methanol/60mM phosphate solution (20/10/70) containing 0.06% (wt/vol) of cetrimide, pH = 7.37; (B) acetonitrile/methanol (75/25) containing 0.06% (wt/vol) of cetrimide. Gradient profile as indicated. Flow rate: 1.04 $\mu$l/min. Column temperature: 31°C. UV detector used at 254 nm. Peaks: as in Fig. 3-20.

ified silica column in which cetyltrimethylammonium bromide (cetrimide) was employed as the detergent. The characteristics of silica columns dynamically modified with cetrimide are similar to those of chemically bonded ODS columns.

## 3.6. Packed Microcapillary Columns

### 3.6.1. Introduction

Packed microcapillary columns have been widely utilized in GC. Such columns are characterized by the ratio of column diameter to particle diameter. Their permeability is between that of densely-packed columns and open-tubular columns. Tsuda and Novótny[54] first applied packed microcapillary columns to LC. The diameter of their columns was around one-tenth of that used in GC.

Due to the good permeability of packed microcapillary columns, large theoretical plate numbers can be achieved using a long column. It has been reported that long packed microcapillary columns can be more easily prepared than densely packed columns, because the particles are packed into the original wide-bore glass tubing and the the packed tubing is then drawn into capillaries.

### 3.6.2. Preparation of Packed Microcapillary Columns

Alumina or silica gel that has been carefully dried is packed into soft glass tubing (0.25–0.6-mm ID and approximately 6-mm OD) under appropriate vibration. Capillaries are subsequently prepared in the glass drawing machine. The diameter of the capillary can be adjusted by changing the drawing ratio and/or the diameter of the original glass tubing. It is better to use soft glass tubes than borosilicate glass tubes, because the former can be drawn at a lower temperature. High-temperature preparation is likely to do damage to the packing material. Hirata et al.[55] prepared various chemically bonded columns in situ by passing the silane reagent through the capillary.

### 3.6.3. Instrumentation

Packed microcapillary LC results in certain instrumental difficulties similar to open-tubular LC. Extremely low flow rates (1 $\mu$l/min or less) were obtained using stream splitting.[54] Stepwise gradient-elution method was also utilized.[56] The sample was loaded using either split injection[54] or the direct injection.[56] UV[54-56] or fluorescence[56] spectrophotometers and electrochemical detection[57] were employed with some modifications.

### 3.6.4. Performance of Packed Microcapillary Columns

Column efficiency is dependent on the ratio of the column diameter to the particle diameter. The optimum ratio is 2–3 for columns using 30-$\mu$m

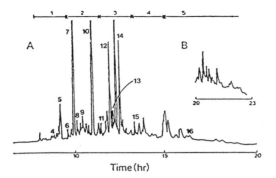

**FIGURE 3-25.** Chromatogram of the aromatic fraction of a coal tar obtained on a packed microcapillary column. Column: 70-$\mu$m ID $\times$ 55-m length, packed with 30-$\mu$m basic alumina with ODS groups. Peaks: 1, benzene; 2, naphthalene; 3, biphenyl; 4, fluorene; 5, phenanthrene; 6, anthracene; 7, fluoranthene; 8, pyrene; 9, triphenylene; 10, benzo(a)anthracene; 11, chrysene; 12, benzo(e)pyrene; 13, perylene; 14, benzo(a)pyrene; 15, dibenzo[ghi]perylene; 16, coronene stepwise gradient was used.

particles and 4–6 for columns using 10-$\mu$m particles. If the ratio exceeds 10, the particles in the column are likely to move downstream.

Stepwise gradient separation of aromatic compounds in coal tar on a 55-m long ODS-bonded basic alumina column of 70 $\mu$m ID is illustrated in Fig. 3-25.[55]

Figure 3-26 shows the separation of phthalates on a 10.3-m long silica gel column with 47-$\mu$m ID.[58] The number of theoretical plates for the last two peaks are 68,000 and 53,000, respectively.

## 3.7. Semi-Microcolumns

### 3.7.1. Introduction

There have been many reports dealing with microsized columns. Several different names have been used to discriminate microsized columns

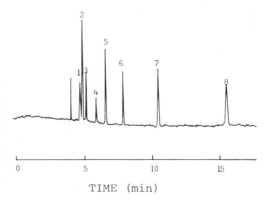

**FIGURE 3-26.** Separation of phthalates on a packed microcapillary column. Column: 47-$\mu$m ID $\times$ 10.3-m length, packed with silica particles. Inlet pressure: 500 kg/cm$^2$. Linear velocity: 4.2 cm/sec. Peaks: 1, didecyl; 2, dinonyl and dioctyl; 3, diheptyl; 4, dicyclohexyl; 5, dibutyl; 6, dipropyl; 7, diethyl; 8, dimethyl phthalate.

**TABLE 3-1**
Summary of Details of Semi-micro-HPLC Columns
and Descriptive Terms Applied to Them

| Description of column | Corresponding range of column ID (mm) |
|---|---|
| Microbore | 0.05–2.8 |
| Narrow-bore | 0.25–1.0 |
| Small-bore | 0.05–2.0 |
| Small-diameter | 2.0 |

from columns having conventional sizes. According to the report of Basey and Oliver,[59] the names listed in Table 3-1 have been used most frequently in the literature. However, it is very difficult to identify the actual dimension of the column internal diameter by these names because they only indicate that the columns are smaller in dimension than conventional columns. If these names are to be used, the actual dimensions must also be specified. Moreover, as mentioned in Chapter 2, the classification of columns by column ID only has little significance; columns should be classified by their volumes. We have defined the columns that have about $\frac{1}{10}$th of the volume of a conventional column (4.6-mm ID × 250-mm length) and give a peak volume of about 10 $\mu$l for the unretained peak as *semi-microcolumns.*

A semi-microcolumn should be distinguished from the real *microcolumns* as a column having less than $\frac{1}{100}$th of the volume of a conventional column. According to the above definition, typical semi-microcolumns are:

    1.0-mm ID × 500-mm length
    1.5-mm ID × 250-mm length
    2.1-mm ID × 100-mm length

A 25–30-mm long column with 4.6-mm ID is now becoming popular and is also classified as a semi-microcolumn. The significance of this classification is that all semi-microcolumns give about 10 $\mu$l peak volume for the unretained peak. For the same peak volume, the HPLC system having the same instrumental band width or the total extracolumn peak volume can be used with the same degree of peak broadening regardless of the column dimensions.

### 3.7.2. Instrumentation

The most significant factor in using semi-microcolumns is that the modification of the conventional HPLC system for compatibility with these columns is minimum. In practice, only replacement of an injector and a detector flow cell with those having proper dimensions, e.g., a microvalve injector with an injection volume of 1–3 $\mu$l and a microflow cell with a volume of 1–2.5 $\mu$l, will be sufficient, as most conventional HPLC pumps can deliver flow rates suitable for semi-microcolumns (100–300 $\mu$l/min) without any problem. Furthermore, semi-microcolumns can be handled without any of the special skills required for handling microcolumns with less than $\frac{1}{100}$ of the volume of a conventional column and they can be used as conventional columns.

### 3.7.3. Sensitivity

As discussed in Chapter 2, the maximum peak concentration of a peak eluted by a semi-microcolumn is ten times higher than that of a peak eluted by a conventional column for a given sample mass under the same chromatographic conditions.[60] Figure 3-27 shows the comparison of the sensitiv-

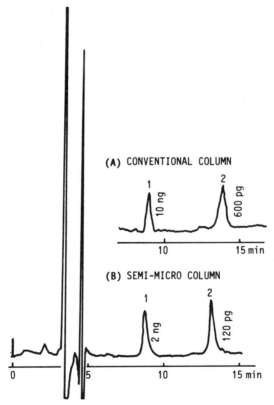

**FIGURE 3-27.** Comparison of sensitivity between semi-micro- and conventional HPLC. Mobile phase: Acetonitrile/water (75/25). Detector: UV at 250 nm, 0.005 AUFS. Peaks: 1, naphthalene; 2, anthracene. (A) Conventional HPLC. Column: 4.6-mm ID × 250-mm length, packed with Fine SIL $C_{18}$. Flow rate: 1.0 ml/min. Detector cell: 1-mm ID, 10-mm path length. (B) Semi-micro-HPLC. Column: 1.5-mm ID × 250-mm length, packed with Fine SIL $C_{18}$. Flow rate: 100 $\mu$l/min. Detector cell: 0.5-mm ID, 5-mm path length.

ities. Because a UV detector cell with 5 mm path was used in the semi-micro-HPLC system, while the conventional cell had a path length of 10 mm, the direct comparison results in a five times higher sensitivity for the semi-micro-HPLC system.

### 3.7.4. Sample Loading Capacity

The maximum amount of sample that can be loaded on a column without losing column efficiency depends on the column volume. Since the ID is in square relation to the column volume, the sample loading capacity of a column rapidly decreases when the diameter is reduced. Figure 3-28 shows the relationship between the amount of sample solute loaded and the column efficiency. A semi-microcolumn (1.5-mm ID × 250-mm length), packed with silica-ODS, was used with benzene ($k' = 4.5$) as the solute. As can be seen, this column can be loaded with as much as 40–50 µg of solute without a significant loss in column efficiency. Considering the fact that the sample amounts are usually from 1 ng to 1 µg for a UV detector, which is most widely used, and from 1 pg to 1 ng for a fluorescence detector, the range of 40–50 µg is large enough to cover almost all demands, while a maximum of 1 mg can be loaded on a conventional column.

### 3.7.5. Solvent Consumption

Since the column volume of a semi-microcolumn is 1/10th of that of a conventional column, the volumetric flow rate required to give the same linear mobile phase velocity in a semi-microcolumn is 1/10th of that needed in a conventional column. Thus, under the same chromatographic conditions with the same analysis time expected, the consumption of the solvent used as the mobile phase is only 10% of that needed in a conventional col-

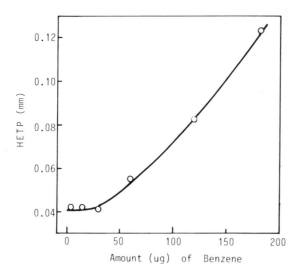

**FIGURE 3-28.** Sample loading capacity of a semi-micro-HPLC column. Column: 1.5-mm ID × 250-mm length, packed with Fine SIL $C_{18}$. Eluent: acetonitrile/water (50/50). Flow rate: 100 µl/min. Solute: benzene ($k' = 4.5$).

umn. In routine analysis, 10% solvent consumption will greatly contribute to the economy of the laboratory. When very expensive and/or environmentally hazardous mobile phase solvents are used, reduced consumption is of great significance. Another aspect of semi-microcolumn analysis, in terms of economy, is that the amount of packing material needed for a semi-microcolumn is also only 10% of that needed for a conventional column.

### 3.7.6. High-Speed Analysis

High-speed HPLC in which the shorter analysis time is achieved without losing column efficiency has recently been tried.[8-12] One way to shorten the analysis time is to use a shorter column. Because the number of theoretical plates is proportional to the column length and poorer resolution is expected by simply shortening length, packing material of smaller particle size is usually used for compensation. The other approach is to increase the linear velocity by increasing the volumetric flow rate. However, this increases the operating pressure and the solvent consumption, and there must be a practical limit for such an increase of flow rate.

Figure 3-29 shows the relationship of the flow rate versus the column

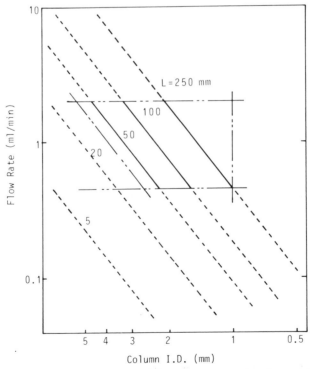

**FIGURE 3-29.** Relationship between flow rate and column ID for high-speed HPLC. Solid lines correspond to a 90% reduction in the analysis time.

ID. The range indicated by the solid lines is for $\frac{1}{10}$th time reduction. If a 4.6-mm ID × 50-mm long column is used with a flow rate of 2 ml/min in a conventional HPLC system, a 90% reduction can be realized. In a semi-micro-HPLC system, a 2.1-mm ID × 100-mm long column with a flow rate of 800 $\mu$l/min can produce similar reduction in the same analysis time. From the point of column volume, the 4.6-mm ID × 50-mm long and 2.1-mm ID × 100-mm long columns give 20% and 10%, respectively, of the peak volume obtained on the conventional column (4.6-mm ID × 250-mm length). Therefore, a dectector cell having 1–2 $\mu$l volume is required for both columns.

### 3.7.7. Preparation of Semi-Microcolumns

Pre-packed columns for semi-micro-HPLC are now commercially available from many manufacturers.[61] However, the selection of column type is still limited as compared with the selection of columns used in conventional HPLC, and the chromatographer must prepare semi-microcolumns with popular materials.

### 3.7.7a. Silica-Based Materials. A semi-microcolumn is packed using the slurry-packing technique just as is a conventional column. For slurry packing, the solvent is first prepared and the packing material is dispersed in it to make the slurry. Then the slurry reservoir, to which an empty column is connected, is filled with this slurry, and the packing solvent is then delivered by a pump to the column via the slurry reservoir. The following three conditions are important in the slurry-packing technique in order to obtain high performance columns:

1. Disperse the packing material well in making the slurry.
2. Keep the slurry stable when the column is being packed.
3. Apply the proper high pressure during packing.

In order to ensure proper dispersion of the packing material, one must prepare a slurry solvent with a high affinity for it. A surfactant is also effective for this purpose. If the slurry solvent is not properly prepared, the packing material will agglomerate or flocculate in the slurry. Table 3-2 shows an

**TABLE 3-2**
An Example of the Composition of a Slurry Solvent

| Component | Concentration (% V/V) |
|---|---|
| Methanol | 10.0 |
| Isopropanol | 5.0 |
| Cyclohexanol | 10.0 |
| Cyclohexane | 4.0 |
| 1,1,1-Trichloroethane | 70.0 |
| Nonipole 40® | 1.0 |

**TABLE 3-3**

Solvents Recommended for Density Balance in
Slurry Packing

| Solvent | Density (g/ml) |
|---|---|
| Diiodomethane | 3.0 |
| 1,1,2,2-Tetrabromoethane | 3.0 |
| Dibromomethane | 2.5 |
| Iodomethane | 2.3 |

example of a slurry solvent used by Kuwata *et al.*[62] for packing silica-ODS (Lichrosorb RP-18) and silica $NH_2$ (Nucleosil $5NH_2$). Methanol and the surfactant Nonipole 40® (nonylphenyl polyethylene glycol "4 moles" ether*) were added to improve dispersion.

The packing material in the slurry may precipitate with time because of the difference between its density and the density of the slurry solvent. The balanced-density slurry technique[63] or the high-viscosity slurry technique[64] are usually used to maintain a stable slurry and to prevent the precipitation of the packing material particles during packing.

The balanced-density technique used a slurry solvent that has the same density as the packing material so that the packing material does not precipitate. In order to make the density of the slurry-solvent equal to that of the packing material, the halogenated high-density solvents listed in Table 3-3 are used. These solvents are toxic and have to be handled carefully. This technique is not used as widely today as it was earlier for this reason.

The aim of the high-viscosity slurry-packing technique is to increase the viscosity of the slurry-solvent, decrease the precipitation rate of the packing material in the slurry, and maintain the slurry in a stable condition during packing. Isopropanol and cyclohexanol (given in Table 3-2) were added for this reason. Other solvents used to increase viscosity include *n*-hexanol, polyethylene glycol 200, ethylene glycol, and glycerol.

It is necessary that the same pressure be applied to every segment of the packing bed of the column and the entire bed must be homogeneous. For this purpose, the following techniques have been examined:

1. Pumping the solvent at a controlled flow rate so that a constant pumping pressure is always applied during packing.
2. Programming the pressure to increase with time.[62] If the packing is performed in the constant-pressure mode, the flow rate will greatly differ between the beginning and the end of the packing procedure. This technique can reduce the flow rate change during the packing.
3. Using a restrictor just after the column is packed.[65] This modification also reduces the changes in the flow rate.

*Sanyo Kasei Kogyo Co., Ltd., 11 Nomoto-cho, Hitotsubashi, Higashiyama-ku, Kyoto, Japan

It is necessary that the pump for packing a semi-microcolumn has a flow rate of several ml/min with a pressure of 500 kg/cm$^2$. Most commercial HPLC pumps are suitable for this purpose.

In packing a column that uses a reciprocating pump operated in the constant-pressure mode instead of a pneumatic operated constant-pressure pump, it should be noted that the attainable pressure is greatly affected by the flow resistance of the packing system, which changes with time during packing procedure because the maximum flow rate of such a pump is generally limited to 10 ml/min. Therefore, the actual pressure often cannot reach the preset value, due to too low flow resistance, until the column is packed with a certain amount of packing material and has a resistance. The pump then runs at a preset pressure, controlling the flow rate automatically.

The viscosity resistance provides the pressure that is applied to the packing material to form a packed bed in the column. The pressure applied to a segment of the packed bed is not directly related to the pump operating pressure or the apparent pressure along the total column length but depends on the flow rate. Therefore, if the flow rate is reduced during the packing process, which normally occurs when a constant-pressure pump is used, the segment of the packed bed at the column top becomes looser than the segment at the bottom. These facts suggest that the constant flow rate mode is more effective when the entire column is homogeneously packed, resulting in a column that has a higher performance. The actual pressure applied in packing greatly affects column performance. For silica-based packing material, a pressure of 500 kg/cm$^2$, which is near the limit of a commercially available HPLC pump, is certainly required. If the pressure is lower, only a column with a low efficiency will be obtained.

### 3.7.7b. Polymer-Based Material.

Porous polymer gel can also be packed by the slurry-packing technique. The following two conditions are very important in porous polymer gel packing:

1. The porous polymer gel is swollen by the solvent but the degree of swelling differs depending on the type or composition of the solvents used. One must therefore use for the slurry and the packing solvents that swell the gel less than the mobile phase that is to be used during analysis. The gel will otherwise shrink in the column during analysis, producing void volumes and resulting in a rapid deterioration of the column.

2. The strength of the porous polymer gel is smaller than that of silica-based packing material (although it depends on the type of the packing material, it is still several tens of kg/cm$^2$ to 200 kg/cm$^2$). In packing, a pressure about 90% of this strength should be applied. If the pressure is not sufficiently high, the peak will show a tailing, while a too high pressure will produce peaks with a leading.

### 3.7.7c. Slurry Reservoir.

The internal volume of the slurry reservoir should be approximately ten times as large as that of the column to be

**TABLE 3-4**

Flow rate range: 10–990 $\mu$l/min (1.0–9.9 ml/min)

| System | Specifications |
|---|---|
| Pump[a] | Flow rate range: 10–990 $\mu$l/min (1.0–9.9 ml/min) Maximum pressure: 500 kg/cm$^2$ Constant flow or constant pressure mode |
| Slurry reservoir | Internal volume: 4 ml |

[a]FAMILIC 300S (JASCO) or similar.

packed. If the slurry reservoir is too large, it will take too long to feed the slurry into the column and packing cannot be carried out well because of the precipitation of the packing material during that time. The end fitting of the column is usually not interchangeable between various commercial products. It is therefore necessary to prepare a slurry reservoir that has a fitting section that matches the empty column to be packed.

Table 3-4 lists the recommended equipment and Table 3-5 lists the rec-

**TABLE 3-5**

Recommended Packing Method for the Preparation of Semi-micro-HPLC Columns by the Slurry-packing Method

| Packing materials | Particle diameter | Column dimensions | Slurry solvent | Slurry | Packing solvent and packing procedure[a] |
|---|---|---|---|---|---|
| File SIL C$_{18}$-10 | | | | | Chloroform (4 ml/min CF → 500 kg/cm$^2$ CP); 15 min |
| Fine SIL C$_8$-10 Fine SIL C$_1$-10 | 10 | 250-mm × 1.5-mm ID | n-Hexanol-chloroform (1:1) | 1g/3ml | Chloroform (4 ml/min CF → 500 kg/cm$^2$ CP); 10 min → methanol (500 kg/cm$^2$ CP), 5 min → methanol/water (500 kg/cm$^2$ CP), 5 min → methanol (500 kg/cm$^2$ CP), 5 min |
| Fine SIL | 5 | | n-Hexanol-di chloromethane (1:1) | | Dichloromethane (2 ml/min CF → 500 kg/cm$^2$ CP), 15 min |
| Fine GEL 110 (polystyrene gel)[b] | 10 | | Methanol | | Methanol (0.5 ml/min CF), 15 min; maximum pressure: 200 kg/cm$^2$ |
| Fine GEL 101[c] (polystyrene gel) | 8 | 2.1-mm ID | Tetrahydrofuran | 1g/8ml | Tetrahydrofuran (0.1 ml/min CF→0.15 ml/min CF), 3 hrs; maximum pressure: 80 kg/cm$^2$ |

[a]CF, constant flow rate method; CP, constant pressure method.
[b]For reversed-phase chromatography.
[c]For size-exclusion chromatography.

ommended packing material and packing procedure for the preparation of semi-microcolumn.

### 3.7.8. Packing Materials

There is no particular restriction on the type of packing material. It is possible to obtain the same properties and performance with a semi-micro-column as with a conventional column by using the same packing material. This ensures that the extended amount of software developed for conventional HPLC can be utilized without modifications for semi-micro-HPLC.

For reference, the list of packing materials compiled by Majors[66] is given in the Appendixes. The polymer-based packing materials have been only rarely used in semi-micro-HPLC. This simply reflects the fact that up to now relatively popular columns, such as silica gel and silica-ODS have been utilized, and it does not mean that the polymer-based packing materials are not suited for semi-micro-HPLC. It is also possible to use porous polymer, ion-exchange resin, and size-exclusion chromatography packing material, as well as others for semi-microcolumns. In fact, Hibi et al.[67] have demonstrated size-exclusion chromatography using semi-microcolumns packed with polymer-based materials.

### References

1. Yang, F.J. *J. Chromatogr.* **1982, 236,** 265.
2. Takeuchi, T.; Ishii, D. *J. Chromatogr.* **1984, 288,** 451.
3. Takeuchi, T.; Ishii, D. *J. Chromatogr.* **1982, 253,** 41.
4. Van der Wal, Sj.; Yang, F. *J. High Resolut. Chromatogr./Chromatogr. Commun.* **1983, 6,** 216.
5. Ishii, D.; Hibi, K.; Asai, K.; Nagaya, M. *J. Chromatogr.* **1978, 152,** 341.
6. Takeuchi, T.; Ishii, D.; Nakanishi, A. *J. Chromatogr.* **1984, 285,** 97.
7. Takeuchi, T.; Ishii, D. *J. Chromatogr.* **1981, 213,** 25.
8. Scott, R.P.W.; Kucera, P.; Munroe, M. *J. Chromatogr.* **1979, 186,** 475.
9. DiCesare, J.L.; Dong, M.W.; Ettre, L.S. *Chromatographia.* **1981, 14,** 257.
10. DiCesare, J.L.; Dong, M.W.; Atwood, J.G. *J. Chromatogr.* **1981, 217,** 369.
11. DiCesare, J.L.; Dong, M.W.; Ettre, L.S. In "Introduction to High-Speed Liquid Chromatography"; Perkin Elmer: Norwalk, CT., **1981,** p 1.
12. Erni, F. *J. Chromatogr.* **1983, 282,** 371.
13. Ishii, D.; Takeuchi, T. *J. Chromatogr.* **1983, 255,** 349.
14. Hirose, A.; Wiesler, D.; Novotny, M. *Chromatographia* **1984, 18,** 239.
15. Takeuchi, T.; Watanabe, Y.; Ishii, D. *High Resolut. Chromatogr./Chromatogr. Commun.* **1980, 4,** 300.
16. Hirata, Y.; Jinno, K. *J. High Resolut. Chromatogr./Chromatogr Commun.* **1983, 6,** 196.
17. Takeuchi, T.; Watanabe, Y.; Matsuoka, K.; Ishii, D. *J. Chromatogr.* **1981, 216,** 196.
18. Takeuchi, T.; Ishii, D.; Saito, M; Hibi, K. *J. Chromatogr.* **1984, 295,** 323.
19. Jinno, K. *High Resolut Chromatogr./Chromatogr. Commun.* **1982, 5,** 364.
20. Fujimoto, C.; Jinno, K. *High Resolut. Chromatogr./Chromatogr. Commun.* **1983, 6,** 374.
21. Golay, M.J.E. *Anal. Chem.* **1957, 29,** 928.
22. Golay, M.J.E. *Anal. Chem.* **1957, 180,** 435.
23. Golay, M.J.E. In "Gas Chromatography 1958"; Desty, D.H., Ed.; Butterworths: London, 1958, p 36.

24. Horváth, C.G.; Preiss, B.A.; Lipsky, S.R. *Anal. Chem.* **1967, 39,** 1422.
25. Nota, G.; Marino, G.; Buonocore, V.; Ballio, A. *J. Chromatogr.* **1970, 46,** 103.
26. Hibi, K.; Ishii, D.; Fujishima, I.; Takeuchi, T.; Nakanishi, T. *J. High Resolut. Chromatogr./ Chromatogr. Commun.* **1978, 1,** 21.
27. Taylor, G. *Proc. R. Soc. London* **1953, 219A,** 186.
28. Hofmann, K.; Halasz, I. *J. Chromatogr.* **1979, 173,** 211.
29. Snyder, L.R.; Dolan, J.W. *J. Chromatogr.* **1979, 185,** 43.
30. Dolan, J.W.; Snyder, L.R. *J. Chromatogr.* **1979, 185,** 57.
31. Tijssen, T. *Sep. Sci. Technol.* **1978, 13,** 681.
32. Hofmann, K.; Halasz, I. *J. Chromatogr.* **1980, 199,** 3.
33. Tsuda, T.; Nomura, K.; Nakagawa, G. *J. Chromatogr.* **1982, 248,** 241.
34. Ishii, D., Takeuchi, T. *J. Chromatogr. Sci.* **1980, 18,** 462.
35. Tijssen, R.; Bleumer, J.P.A.; Smit, A.L.C.; Van Kreveld, M.E. *J. Chromatogr.* **1981, 218,** 137.
36. Krejci, M.; Tesarik, K.; Rusek, M.; Pajurek, M. *J. Chromatogr.* **1981, 218,** 167.
37. Tsuda, T.; Tsuoboi, K.; Nakagawa, G. *J. Chromatogr* **1981, 214,** 283.
38. Jorgenson, J.W.; Guthrie, E.J. *J. Chromatogr.* **1983, 255,** 335.
39. Knox, J.H.; Gilbert, M.T. *J. Chromatogr.* **1979, 186,** 405.
40. Bristow, P.A.; Knox, J.H. *Chromatographia* **1977, 10,** 279.
41. Knox, J.H. *J. Chromatogr. Sci.* **1980, 18,** 453.
42. Hibi, K.; Tsuda, T.; Takeuchi, T.; Nakanishi, T.; Ishii, D. *J. Chromatogr.* **1979, 175,** 105.
43. Ishii, D.; Takeuchi, T. In "Advances in Chromatography"; Giddings, J.S.; Grushka, E.; Cazes, J.; Brown, P.R. Eds.; Marcel Dekker: New York, 1983; Vol. 21, p 131.
44. Takeuchi, T.; Ishii, D. *J. Chromatogr* **1982, 240,** 51.
45. Ishii, D.; Tsuda, T.; Takeuchi, T. *J. Chromatogr* **1979, 185,** 73.
46. Ishii, D.; Takeuchi, T. In "Reviews in Analytical Chemistry"; West, T.S., Ed.; Freund Publishing House: Aberdeen, Scotland, 1982; Vol. IV, No. 2, p 87.
47. Yang, F.J. *J. High Resolut. Chromatogr./Chromatogr. Commun.* **1980, 3,** 589.
48. Ishii, D.; Takeuchi, T. *J. Chromatogr. Sci.* **1984, 22,** 400.
49. Takeuchi, T.; Ishii, D. *J. Chromatogr.* **1983, 279,** 439.
50. Ishii, D.; Takeuchi, T. *J. Chromatogr.* **1981, 218,** 189.
51. Takeuchi, T.; Matsuoka, K.; Watanabe, Y.; Ishii, D. *J. Chromatogr.* **1980, 192** 127.
52. Takeuchi, T.; Kitamura, H.; Spitzer, T.; Ishii, D. *J. High Resolut. Chromatogr./Chromatogr. Commun.* **1983, 6,** 666.
53. Takeuchi, T.; Ishii, D.; Nakanishi, A. *J. Chromatogr.* **1983, 281,** 73.
54. Tsuda, T.; Novótny, M. *Anal. Chem.* **1978, 50,** 271.
55. Hirata, Y.; Novótny, M.; Tsuda, T.; Ishii, D. *Anal. Chem.* **1979, 51,** 1807.
56. Hirata, Y.; Novótny, M. *J. Chromatogr.* **1979, 186,** 521.
57. Hirata, Y.; Lin, P.T.; Novótny, M.; Wightman, R.W. *J. Chromatogr.* **1980, 181,** 287.
58. Tsuda, T.; Tanaka, I.; Nakagawa, G. *Anal. Chem.* **1984, 56,** 1249.
59. Basey, A.; Oliver, R.W.A. *J. Chromatogr.* **1982, 251,** 265–268.
60. Saito, M.; Wada, A.; Hibi, K.; Takahashi, M. *Industrial Research/Development* **1983, 25,** 102–106.
61. *"International Chromatography Guide", J. of Chromatogr. Sci.,* **24,** 1G (1986).
62. Kuwata, K.; Uebori, M.; Yamazaki, Y. *J. Chromatogr* **1981, 211,** 378–382.
63. Kirkland, J.J. *J. Chromatogr. Sci.* **1971, 9,** 206.
64. Endele, R.; Halasz, I.; Unger, K. *J. Chromatogr.* **1974, 99,** 377.
65. Halasz, I.; Maldener, G. *Anal. Chem.* **1983, 55,** 1842–1847.
66. Majors, R.E. *J. Chromatogr. Sci.* **1980, 18,** 488–511.
67. Hibi, K.; Wada, A.; Saito, M.; Arita, M. Pittsburgh Conference of Analytical Chemists, Atlantic City, NJ, March, 1984; Abstr. **553.**

# 4

# Detection Systems

## M. Saito, K. Hibi, and M. Goto

### 4.1. Introduction

In HPLC, there is an inherent problem in the selection of the proper detection systems, as the physical properties of both the mobile phase and the sample solutes are quite similar. Snyder and Kirkland[1] classified the approaches to detection problems as

1. Differential measurement of a bulk or general property of both sample and solvent,
2. Measurement of a sample property that is not possessed by the mobile phase, and
3. Detection after eliminating the mobile phase.

The first type of detection system is universal and general purpose and includes a refractive index detector. In the early stages of HPLC, the refractive index detector was often used in connection with a UV absorption detector. At present, however, refractive index detectors have become less important as general purpose detectors, and are used mainly for polymer analysis, i.e., size-exclusion chromatography. This detector has hardly been used in micro-HPLC, partly because it is not easy to make a small-volume cell, and heat-exchanger which is necessary for the detector to obtain a stable baseline, and partly because there are many other detection systems that are more attractive than a refractive index detector.

The second type of detection system, now most widely used, includes UV absorption detectors (fixed, variable, and multi-wavelength detectors), fluorescence detectors, light-scattering detectors, and electrochemical detectors. Post-column derivatization detection systems can also be classified as

**M. Saito and K. Hibi** Japan Spectroscopic Company, Ltd., Hachioji City, Tokyo 192, Japan.    **M. Goto** Research Center for Resource and Energy Conservation, Nagoya University, Chikusa-ku, Nagoya 464, Japan.

the second type. The systems incorporate a reaction system and a detection system for reaction products. (See Chapter 6.)

The last type of detection system has been proposed for LC-FID (flame ionization detector), LC-MS, and LC-IR spectrometers. The LC-IR and LC-MS will be discussed in detail in the section on hyphenated LC in Chapter 5 (Sections 5-1 and 5-2).

Advantages of micro-HPLC, such as high solute concentrations and low flow rate, offer significant differences between the sample solutes and the mobile phase, favoring easier detection with any detection method. Here, the UV absorption detector, fluorescence detector, and voltammetric detector in connection with micro- and semi-microcolumns will be discussed.

### 4.2. UV Absorption Detectors

Among the variety of HPLC detectors, the UV absorption detectors have been most widely used. For many chemical compounds, such a detector offers high sensitivity, good stability, and good linear dynamic range and permits a wide range of applications. In micro-HPLC, a UV absorption was first used as the detection system. Ishii and co-workers[2] showed how a commercial variable-UV detector could be modified for use with one-hundredth downscaled micro-HPLC system. Scott and Kucera[3] also described the modification of the flow cell of a commercial fixed-wavelength UV detector to meet the requirements of one-tenth downscaled microscale columns. At present, UV absorption detectors that are compatible with small-volume columns are commercially available.[4,5]

The commercial UV absorption detectors can be classified into three groups:

1. Fixed wavelength
2. Variable wavelength
3. Multi-wavelength

The first type of detectors are simple, inexpensive, and easy to modify for the use in micro-HPLC. Low-pressure mercury lamps are used as light sources for this type of detector. Such a detector has very high sensitivity [noise level: $2 \times 10^{-5}$ absorbance units $(AU)$] owing to the very strong line emission at 254 nm from the lamp.

The second type of detectors are in the moderate price range and easy to modify for the use in micro-HPLC. They are now becoming more popular because they offer the greater versatility and the convenience of detection at the absorption maximum of the sample solute. Deuterium and tungsten lamps are used as light sources for this type of detector. The deuterium lamp covers the UV wavelength range from 190 to 350 nm (sometimes to 600 nm with a UV cut filter) and the tungsten lamp covers the visible wavelength range from 350 to 700 nm. The noise level of such a detector is generally 5 $\times 10^{-5}$ AU (about 2.5 times higher than that of a fixed-wavelength detector). In practice, a variable-wavelength detector often offers higher sensitivity than a fixed-wavelength detector when the wavelength is tuned at the absorption maximum of the sample solute.

The last type of detector is the most sophisticated and is used with data-processing systems, permitting presentation of three-dimensional chromatograms, contour plots, and ratio chromatograms. However, the modification of the flow cell is generally difficult because of the sophisticated optics.

### 4.2.1. Flow Cells

Instrumental requirements for UV detectors in microscale HPLC have been discussed in detail in Chapter 2. Some of the practical aspects of performing microscale HPLC with a commercial UV detector will be dealt with in this chapter. To select a UV detector for micro-HPLC, it is quite important to ensure that the detector be equipped with a cassette-type flow cell, so the chromatographer can use different types of flow cells, depending on the peak volume obtained on the column to be used.

Table 4-1 lists the volumes of flow cells needed for the different types of microscale columns. Flow cells of less than 0.3 $\mu$l volume are generally not commercially available, and the chromatographer has to prepare such cells. Figures 4-1 and 4-2 show examples of such cells, as prepared by Takeuchi and co-workers[6] and Yang[7] for use with microcolumns having less than 1/100th of the volume of a conventional column. These cells were installed in the position of the hole of the original cell, which functions as the exit slit for the monochromator of the detector.

Figure 4-3 shows the configuration of a conventional 8-$\mu$l $\times$ 10-mm path length cell and a microscale 1-$\mu$l $\times$ 5-mm path length cell. System variances that include these cells are plotted for various flow rates in Fig. 4-4. The measurements were carried out by connecting these cells directly to an injector with a 1-$\mu$l loop.[8] As shown, there are significant differences between the variances obtained by the two cells.

Figure 4-5 compares two chromatograms. Chromatogram A was obtained by using the 1-$\mu$l cell and chromatogram B was obtained by using

**TABLE 4-1**
Various Cell Volumes for UV Detectors Used in Micro-HPLC

| | Column dimensions | | | Flow cell |
| Type of HPLC | Length (cm) | ID (mm) | Column tube material | volume ($\mu$l) |
|---|---|---|---|---|
| Open-tubular LC | 2,000 | 0.03 | Glass | 0.03–0.05 |
| Packed microcapillary LC | 1,000–2,000 | 0.03–0.1 | Glass | 0.02 |
| Micro-HPLC | 15 | 0.5 | PTFE | 0.3 |
| Micro-HPLC | 10–30 | 0.2–0.35 | Fused-silica | 0.05 |
| Semi-micro-HPLC | 25 | 1.5 | Stainless-steel | 1.0 |
| Conventional HPLC[a] | 25 | 4.6 | Stainless-steel | 8.0 |

[a]Given for comparison.

OUTLET

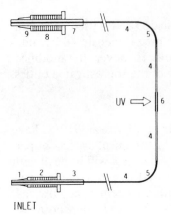

INLET

**FIGURE 4-1.** Schematic diagram of a microflow cell prepared by Takeuchi et al.[6] The cell withstands 200 kg/cm$^2$ pressure and can be used not only for micro-HPLC, but also for microcolumn supercritical chromatography. 1, $\frac{1}{32}$-inch ferrule; 2, $\frac{1}{32}$-inch compression screw; 3, 0.25-mm ID × 0.8-mm OD stainless-steel tubing; 4, 55-$\mu$m ID × 0.24-mm OD fused-silica tubing; 5, 0.33-mm ID × 0.63-mm OD stainless-steel tubing; 6, 0.26-mm ID × 0.37-mm OD fused-silica tubing; 7, 0.3-mm ID × $\frac{1}{16}$-inch OD stainless-steel tubing; 8, $\frac{1}{16}$-inch compression screw; 9, $\frac{1}{16}$-inch ferrule. (Reprinted from Takeuchi *et al.*[6] with permission.)

the 8-$\mu$l cell. The peak broadening in the 8-$\mu$l cell is so large that resolutions between peaks 2, 3, and 4 are seriously damaged.

It has been emphasized many times in previous chapters that even the smallest dead volume must be eliminated in micro-HPLC. Even if a detector having a flow cell with a proper volume is used, careless connection with improper tubing often destroys the overall performance of the system. Figure 4-6 shows a normal chromatogram (A) and a deteriorated chromatogram (B) that exhibit peaks with severe tailing due to bad connections between the injector, the column, and the detector cell.

### 4.2.2. Sensitivity

As discussed in Section 2-1 and in Fig. 2-3 and Table 2-1, high sample solute concentration is one of the most important advantages of using

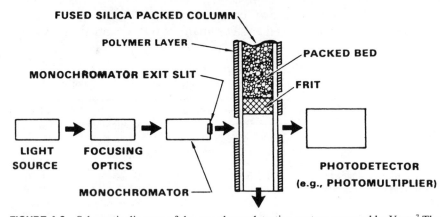

**FIGURE 4-2.** Schematic diagram of the on-column detection system prepared by Yang.[7] The end of a narrow-bore fused-silica column serves as the flow cell. (Reprinted from Yang[7] with permission.)

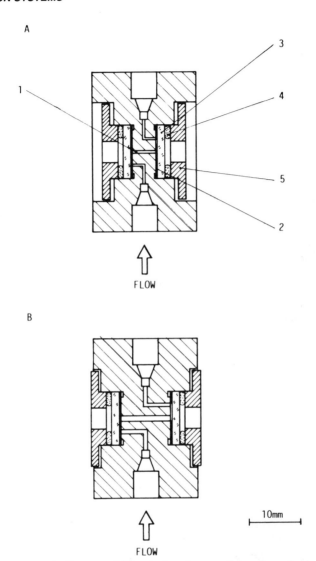

**FIGURE 4-3.** Schematic diagrams of (A) a 1-μl × 5-mm path length semi-micro-HPLC cell and (B) a 8-μl × 10-mm path length conventional cell, both drawn in the same scale and intended for the JASCO UVIDEC-100-V UV detector. 1, 0.5-mm ID × 5-mm length cell cavity; 2, PTFE gasket with a slit for the solvent path; 3, quartz window; 4, O-ring; 5, compression flange. (Courtesy of JASCO.)

microscale columns. The output of a UV absorption detector at a certain wavelength is represented in terms of absorbance units (*AU*) by Beer's law.

$$AU = \epsilon_\lambda l_{dc} C_{sol} \tag{1}$$

where $\epsilon_\lambda$ is the molar absorptivity at the wavelength ($\lambda$) of the sample solute, $l_{dc}$ is the detector cell path length (cm), and $C_{sol}$ is the solute concentration

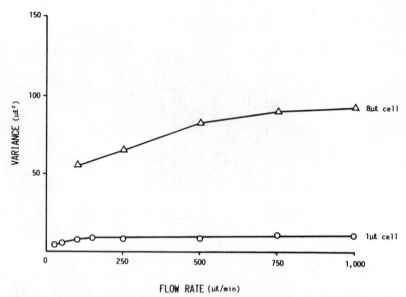

**FIGURE 4-4.** System variances with (O) 1- and (△) 8-μl cells as a function of the flow rate. The measurements were performed by connecting each cell directly to a microvalve injector with a 1-μl loop. The variances were calculated by measuring the peak width ($2\sigma$) at 0.607 of the peak height.

(moles/l). Equation 1 can be rewritten by using equation 5 from Chapter 2. The maximum absorbance ($AU_{p(max)}$) of the peak is then expressed as

$$AU_{p(max)} = 1600 \frac{\epsilon l_{dc} m_s}{V_p (MW)} \qquad (2)$$

where $MW$ is the molecular weight of the sample solute, $m_s$ is the sample mass (g), and $V_p$ is the peak volume (ml). Therefore, the maximum absorbance output for a given sample solute is proportional to the detector path length ($l_{dc}$) and inversely proportional to the peak volume ($V_p$).

Figure 4-7 compares the sensitivity of conventional HPLC and one-tenth downscaled, i.e., semi-micro-HPLC.[9] Chromatogram A was obtained by using a conventional column and a UV detector with the 8-μl cell (path length: 10 mm), while chromatogram B was obtained by using a semi-micro-column and the same UV detector with the 1-μl cell (path length: 5 mm). The linear velocities were kept the same for both columns in order to maintain the same retention times. As the peak volume eluting from a semi-microcolumn is one-tenth of the peak volume eluting from a conventional column, the maximum absorbance of the peak in semi-micro-HPLC should be ten times higher for the same path length. In this case, the path length of the 1-μl cell is one half that of the 8-μl cell. Therefore, one-fifth of the sample amount used in the conventional column gave the same absorbance for the peaks eluting from the semi-microcolumn, as shown in Fig. 4-7. Since the

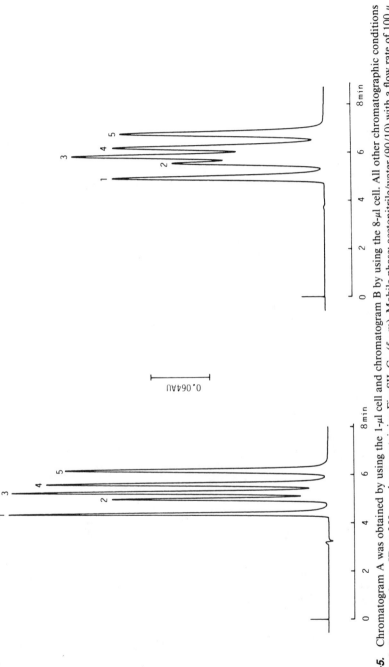

**FIGURE 4-5.** Chromatogram A was obtained by using the 1-µl cell and chromatogram B by using the 8-µl cell. All other chromatographic conditions were identical. Column: 1.5-mm ID × 250-mm long, containing FineSIL C$_{18}$ (5 µm). Mobile phase: acetonitrile/water (90/10) with a flow rate of 100 µ l/min. Sample volume: 1 µl. Detector used at 250 nm. Peaks: 1, benzene (1.0%); 2, naphthalene (0.05%); 3, biphenyl (0.01%); 4, fluorene (0.01%); 5, anthracene (0.001%) in acetonitrile.

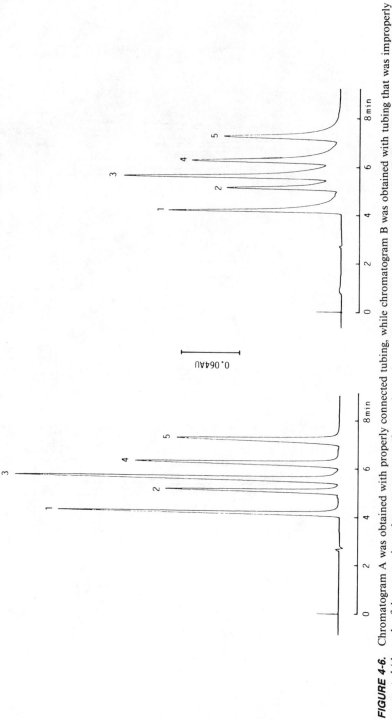

**FIGURE 4-6.** Chromatogram A was obtained with properly connected tubing, while chromatogram B was obtained with tubing that was improperly connected (the ends of the connecting tubes were not seated properly on the bottoms of the holes of both column end-fittings). The 1-$\mu$l cell and the 1-$\mu$l loop injector were used. Other chromatographic conditions were the same as in Fig. 4-5.

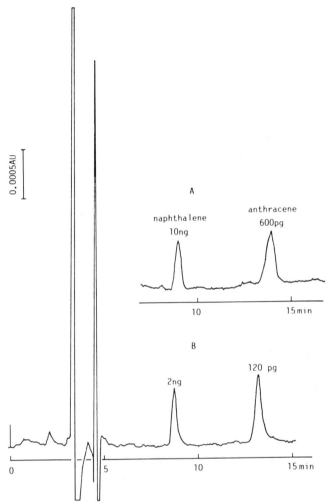

**FIGURE 4-7.** Comparison of the sensitivity obtained in conventional HPLC and semi-micro-HPLC. Chromatogram A was obtained by using a conventional column (4.6-mm ID × 250-mm length), containing FineSIL C$_{18}$ (10$\mu$m) and the 8-$\mu$l cell (path length: 10 mm). Chromatogram B was obtained by using a semi-microcolumn (1.5-mm ID × 250-mm length), containing the same material and the same detector with a 1-$\mu$l cell (path length: 5 mm). Mobile phase: acetonitrile/water (75/25) with flow rates of 1.0 ml/min for A and 100$\mu$l/min for B.

optical apertures of both cells are the same, the noise levels of these cells are very similar, and the sensitivity was thus incrased by a factor of five.

### 4.2.3. Time Constant

As discussed in Section 2.2 the detector time constant is a critical factor in high-speed analysis. Figure 4-8 shows chromatograms obtained by the same UV detector with different settings of the time constant. Chromatogram A

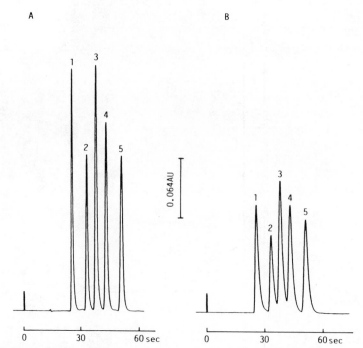

**FIGURE 4-8.** Influence of the time constant on peak shape. Chromatogram A was obtained with 0.05-sec time constant, while chromatogram B was obtained with 1.0-sec time constant. Column: 4.6-mm ID × 50-mm long, containing FineSIL $C_{18}$ (3$\mu$m). Mobile phase: acetonitrile/water (85/15) with 2 ml/min flow rate. Injection volume: 3$\mu$l. Sample: same as for Fig. 4-5.

was obtained with 0.05-sec time constant, and even though five peaks were eluted within the time period of 60 sec, excellent peak shapes are obtained. On the other hand, chromatogram B, which was obtained with a 1.0-sec time constant, was seriously distorted by the slow time constant, which would not allow quantitative analysis. A 4.6-mm ID × 50-mm long column packed with 3-$\mu$m ODS was used with a flow rate of 2.0 ml/min.

### 4.3. Fluorescence Detectors

Fluorescence detectors are very sensitive and they are selective means of detection for compounds that emit fluorescence by excitation with UV radiation. The sensitivity is high enough to detect picograms and even femtograms of certain compounds present in the flow cell. Both the high sensitivity and the selectivity of a fluorescence detector are inherently related to the detection principles.

In absorption detection, the signal to be detected is a small change on a high background, whereas in fluorescence detection, the signal, though it is weak, comes out of a dark background. This allows the operation of a detection device, such as a photomultiplier, with substantially high-gain amplification. Thus, the main noise sources of a fluorescence detector are the photomultiplier dark-current noise, pre-amplifier noise, background fluorescence, which often comes from the solvent, and stray light, which originates from the excitation beam, often scattering at the flow cell.

Fluorescent compounds are excited by absorption of UV radiation at certain wavelengths and they then emit fluorescent radiation at longer wavelengths, as shown in Fig. 4-9. The excitation spectrum is obtained by scanning the excitation wavelength while monitoring the emitted fluorescence, usually at its maximum emission wavelength. The excitation spectrum shows the dependency of the fluorescent intensity on the excitation wavelength and it is generally similar to the absorption spectrum of the compound. The emission spectrum is obtained by scanning the monitoring wavelength of the fluorescence emission, while the excitation wavelength is usually fixed at the excitation or absorption maximum. Each compound has

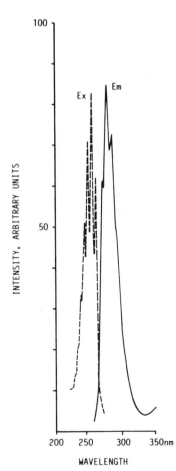

**FIGURE 4-9.** The excitation (Ex) and emission (Em) spectra of benzene. Sample: 100-ppm benzene in methanol. Instrument: JASCO FP-550A spectrofluorometer. Excitation spectra measuring conditions: excitation wavelength, 220–300 nm; emission wavelength, 280 nm; excitation slit width, 3 nm; emission slit width, 10 nm. Emission spectra measuring conditions: excitation wavelength, 250 nm; emission wavelength, 250–300 nm; excitation slit width, 5 nm; emission slit width, 5 nm.

its own characteristic excitation and emission spectra, which is dependent on the molecular structure. Therefore, the chromatographer can accurately aim the target sample component by tuning the excitation wavelength and emission wavelength to their maxima. In this way, the sample compound in the column effluent is detected through two windows, the excitation wavelength and the emission wavelength. Consequently, the selectivity of the compound is significantly increased. The high selectivity of a fluorescence detector allows specific detection of compounds of interest in complex multicomponent samples, such as biological fluids, foods, pharmaceuticals, and environmental samples. In addition, pre- and post-column derivatization techniques permit the detection of nonfluorescent compounds by using a fluorogenic reagent that only reacts with a specific group of compounds and enhances the selectivity. This will be described in detail in Chapter 6. Thus, applications of fluorescence detection are closely related to pre- and/ or post-column derivatization. Kucera and Umagat[10] reported a microscale post-column fluorescence derivatization system for amine analysis. Takeuchi et al.[11] reported a post-column fluorescence derivatization system for bile acid analysis, using a reactor column packed with immobilized enzyme material.

### 4.3.1. Detection Systems

The detection system used in fluorescence detection is substantially different from that used in absorption detection. In fluorescence detection, UV radiation from the light source illuminates the flow cell and excites the solute molecules present. The solute molecules then emit an equal amount of fluorescence radiation in all directions. In principle, the radiation can be observed from any angle. In practice, the fluorescence is usually detected at a 90-degree angle to the excitation beam. In this way, the excitation light does not stray into the fluorescent radiation, resulting in low background and noise. A unique cell design for efficient collection of the fluorescence from a small volume cell was reported by Schoeffel and Sonnenschein.[12] In their emission collection system, a thin flow cell is illuminated by an excitation beam that originates from the back side of a concave mirror and passes through a transparent hole in the center of the mirror, which is placed around the cell. The emission from the surface of the cell is collected by the same concave mirror.[13,14]

Since fluorescent radiation originates in excited solute molecules, the intensity of the fluorescence is fundamentally proportional to the number of illuminated molecules and, consequently, to the volume of the solution illuminated by the excitation light. However, the fluorescent intensity to be observed cannot be easily calculated, for three reasons. First, because the intensity of the excitation light is exponentially decreased by absorption as the beam travels in the solution; second, because the emitted fluorescence itself is absorbed in the solution; and, finally, because the efficiency of emission collection greatly affects the detector output. Therefore, a fluorescence intensity or an output of a fluorescence detector is always represented by an arbitrary or a relative unit and, as opposed to absorption detection, there is no absolute indication of fluorescence detection.

The detection systems can be classified into three groups:

1. Filter/filter
2. monochromator/filter
3. monochromator/monochromator (spectrofluorometer)

The first type of system (filter/filter) is inexpensive and suitable for routine analysis, such as amino-acid analysis with OPA post-column derivatization (see Chapter 6). Mercury lamps are widely used as the excitation light sources. However, the available excitation wavelengths are quite limited because these lamps have only discrete line emissions. Among them, the 254- and 365-nm lines are most frequently used. Other lines are less frequently utilized due to comparably weak radiant energy and a lack of usable optical filters.

The second type of system (monochromator/filter) is more selective and versatile because the excitation wavelength can be selected. For this type of detector, excitation light sources having continuous emission spectra, such as xeon and deuterium lamps, are used.

The third type of system (monochromator/monochromator) are the most selective and versatile, because both the excitation and emission wavelengths are tunable. In some cases, a spectrofluorometer permits the detection of a specific component in a fused chromatographic peak.[15] Some of these spectrofluorometers can also be used as filter/filter and monochromator/filter types, by utilizing zeroth order light and appropriate filters. The zeroth order light is obtained by turning a grating to the angle where the grating functions as a simple reflecting mirror. For a dedicated HPLC spectrofluorometric detector, spectral band widths for both monochromators are wider (usually 10–20 nm), in order to obtain higher sensitivity, than those in general-purpose spectrofluorometers.

## 4.3.2. Sensitivity versus Flow Cell Volume

The permitted maximum cell volume discussed in Chapter 2 is also valid for a fluorescence detector, but the problem is not as simple as it is for a UV detector. In absorption detection, the sensitivity can basically be maintained as long as the path length holds the same value. In fluorescence detection, it is fundamentally proportional to the illuminated volume of the solution. Therefore, reduction of the cell volume involves a compromise between the sensitivity and the contribution of peak broadening, and it may not be an easy tradeoff. If the detector cell volume is reduced in correspondence with the reduction of the peak volume, the illuminated volume of the sample solute is also reduced by the same factor, resulting in a reduction of the detector response for a given mass of the solute even though the solute concentration is substantially higher than when a conventional column is used.

Since one of the most advantageous features of a fluorescence detector is the high sensitivity, the cell volume should be determined in practice to suit the volume of the peak of interest. For semi-microcolumns, 3- to 6-$\mu$l cells can still be used for detecting moderately retained peaks (k' = 3 − 6) without serious peak broadening and with increased sensitivity due to the high peak concentration. On the other hand, due to the reduction of the illuminated volume, the use of 1- $\mu$l cell would sacrifice what is gained from using semi-microcolumns.

Figure 4-10 compares chromatograms obtained by using various cells. The same amounts of the sample components were used for all the chromatograms. The same semi-microcolumn was used for A to C, and a conventional column was used for D, for reference. Chromatograms A and B, which were obtained by using cells of 1.5- and 3- $\mu$l illuminated volumes, respectively, exhibit good peak shapes and resolution. However, chromatogram C, obtained by using a conventional cell with 7- $\mu$l illuminated volume, shows considerable loss in resolution. Although the noise level increases a little as the illuminated volume is enlarged, the peak heights are also increased and the signal-to-noise ratio is improved in proportion to the volume. Chromatogram D, obtained by using the conventional column and a conventional 7- $\mu$l cell, shows peak heights and baseline noise similar to those of chromatogram A.

Figure 4-11 demonstrates the excellent selectivity of a spectrofluorometric detector. The same amount of the same sample was used for all the chromatograms. Both the excitation and emission wavelengths were tuned to peak 6 (anthracene). Even though this peak appeared relatively small in Fig. 4-10, it is now very large with good signal-to-noise ratio, while other peaks can hardly be seen in all the chromatograms shown in Fig. 4-11.

Table 4-2 lists the minimum detectability values for anthracene for the four chromatograms shown in Fig. 4-11. As shown, the minimum detectable concentration is proportional to the illuminated volume, while the minimum detectable injected amount is also proportional to the illuminated volume, as long as the peak broadening is insignificant. Accordingly, the sensitivity measured with the cell have 7- $\mu$l illuminated volume did not increase as much as expected from (or by?) the ratio of the volumes. The minimum detectable amount measured using the conventional column and the 7- $\mu$l cell is poorer than the amount measured using the semi-microcolumn and the 1.5- $\mu$l cell, even though the 7- $\mu$l cell has 4 times higher sensitivity for concentration. In the case of the 7- $\mu$l cell used with the semi-microcolumn, an injected amount as low as 0.14 pg could be detected, which represents about 5 times better sensitivity than that measured with the conventional column and the same cell. Naturally, it should be noted that the use of the 7- $\mu$l cell causes considerable peak broadening if the peak is eluted

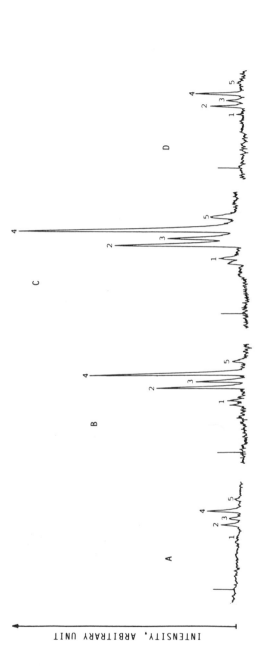

**FIGURE 4-10.** Comparison of chromatograms by using cells with different illuminated volumes for the same sample amount. Chromatograms A, B, C, and D were obtained by using cells with 1.5-, 3.0-, 7.0-, and 7.0-$\mu l$ illuminated volumes, respectively. A 1.5-mm ID × 250-mm long semi-microcolumn packed with FineSIL $C_{18}$ (10$\mu$m) was used for A through C, while a 4.6 mm ID × 250 mm long conventional column packed with FinePack SIL $C_{18}$ (10$\mu$m) was used for D. Injection volume: 1$\mu$l. Detector: JASCO Model FP-210 spectrofluorometric detector. Excitation: 252 nm. Emission: 360 nm. Peaks: 1, benzene (10 ppm); 2, naphthalene (500 ppb); 3, biphenyl (100 ppb); 4, fluorene (100 ppb); 5, anthracene (10 ppb) in acetonitrile.

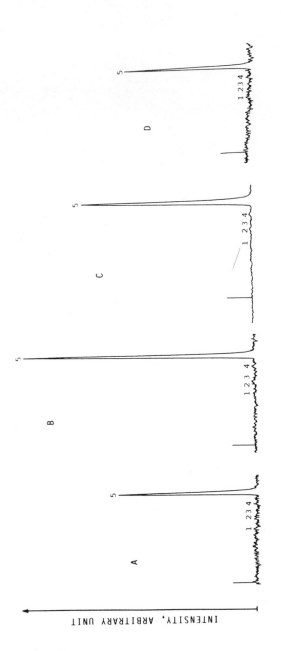

**FIGURE 4-11.** Demonstration of selectivity of the spectrofluorometric detector. The sample, columns, detector cells, and injection volumes are the same as in Fig. 4-10. Both the excitation (255 nm) and the emission (395 nm) wavelengths were tuned to anthracene.

**TABLE 4-2**

Cell Volume versus Sensitivity in Fluorescence Detection[a]

| Chromatogram | Column[b] | Cell volume (illuminated volume) ($\mu$l) | Minimum detectable amount of anthracene[c] (S/N = 2) (pg) | Minimum detectable concentration of anthracene (S/N = 2) (ppb) |
|---|---|---|---|---|
| A | SM | 6.0 (1.5) | 0.50 | 0.02 |
| B | SM | 6.0 (3.0) | 0.25 | 0.01 |
| C | SM | 15.0 (7.0) | 0.14 | 0.005 |
| D | C | 15.0 (7.0) | 0.65 | 0.005 |

[a]Calculated from the chromatograms given in Fig. 4-11.
[b]SM, semi-microcolumn (1.5-mm ID $\times$ 250 mm length, packed with 10$\mu$m ODS); C, conventional column (4.6-mm ID $\times$ 250-mm length, packed with 10$\mu$m ODS).
[c]Injected amount.

too early, i.e., has a small k'. However, this is unlikely to occur in reversed-phase separation of this type of compound.

### 4.3.3. Time Constant

The problem with the detector time constant is basically the same for a fluorescence detector as for other detectors. However, it is worthwhile to mention that, by its nature, the noise from a fluorescence detector is very much like a white noise, which contains all frequencies equally, from low to high. For this type of noise, a low-pass filter is very effective. A time constant faster than necessary only results in excessive noise, reducing the sensitivity. On the other hand, a time constant slower than necessary results in distortion of the peak shape and peak broadening, reducing quantitative accuracy. Therefore, the time constant must be carefully selected with reference to the time width of the peak of interest.

Figure 4-12 shows chromatograms obtained by the same fluorescence detector with different settings of the time constant. Chromatogram B, with a 2-sec time constant, is the same as chromatogram A in Fig. 4-10, while chromatogram A in the present figure was obtained under the same conditions except for the time constant (0.2 sec). A time constant of 2 sec distorts peak shape by 5–10%, while increasing the time constant from 0.2 to 2 sec significantly improves signal-to-noise ratio by more than a factor of three. Chromatograms C and D permit another comparison. These were obtained by using a 4.6 mm ID $\times$ 50 mm long column, which had 20% of the volume of a conventional column and gave a comparably short analysis time. Chromatogram C, with 0.2-sec time constant, shows fairly good peak shapes and

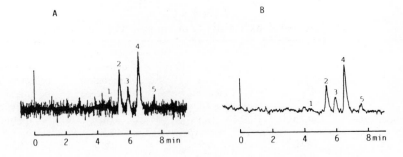

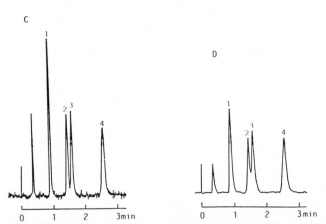

**FIGURE 4-12.** Influence of the time constant on peak shape and sensitivity. Time constant was 0.2 sec for chromatograms A and C and 2 sec for chromatograms B and D. Columns: (A) and (B), 1.5-mm ID × 250-mm long, containing FineSIL $C_{18}$ (10μm); (C) and (D), 4.6 mm ID × 50 mm, containing FineSIL $C_{18}$ (3μm). The sample used for A and B was the same as in Figs. 4-10 and 4-11; the sample used for chromatograms C and D consisted of 1, α-tocopherol; 2, β-tocopherol; 3, γ-tocopherol; 4, δ-tocopherol.

resolutions, even for peaks having retention times of less than 2 min. Chromatogram D peaks, obtained with a 2-sec time constant and retention times of less than 2 min, were seriously broadened. However, a peak eluted at about 2.5 min showed only a little broadening, while the signal-to-noise ratio was significantly improved.

Thus, the time constant should be selected to obtain the maximum sensitivity with tolerable broadening of the peak of interest.

## 4.4. Voltammetric Detectors

As discussed earlier, the primary advantage of microcolumns and semi-microcolumns is in the low solvent consumption and higher mass sensitivity achieved due to the decrease in the mobile-phase flow rate. Voltammetric detectors have a real advantage for such columns, since the low flow rates actually favor electrode efficiency and, even when the cell volume is reduced, sensitivity does not have to be sacrificed.

The miniaturized voltammetric detectors suitable for HPLC using microcolumns and semi-microcolumns are discussed in the following sections.

### 4.4.1. Miniaturized Single Voltammetric Detectors

The miniaturized voltammetric detectors with a single working electrode can be divided into two types: thin-layer cells[16,17] and tubular cells.[18,19]

In the thin-layer cell, built by Goto and co-workers,[16] the cell cavity is constructed of two fluorocarbon resin blocks separated by a 20–45-$\mu$m thick and 0.5–2-mm wide PTFE sheet. A working electrode is made with glassy carbon disks of 3 mm diameter and contained in one of the blocks. The silver-silver chloride reference electrode is held in a cylindrical hole in the other block. A stainless-steel tube serves both as the counter electrode and the exit line. The cell volume of such cells is about 0.06–0.3 $\mu$l.

The cell was tested for the determination of aminophenol isomers separated on a microcolumn (0.5-mm ID $\times$ 147-mm length) packed with silica-ODS at a flow rate of 8.3 $\mu$l/min. The detection limits for isomers were about 10 pg and the linear dynamic range was about $10^3$. The cell was utilized for the determination of bile acids separated on a microcolumn at 8.3 $\mu$l/min with 3$\alpha$-hydroxysteroid dehydrogenase (3$\alpha$-HSD) post-column derivatization.[20] The microcolumn consisted of a 0.5 mmID $\times$ 114 mm long PTFE tube packed with 5 $\mu$m silica-ODS particles. The post-column was made of a 0.5 mmID $\times$ 20 mm long PTFE tube packed with HSD-immobilized controlled-porosity glass beads (200–400 mesh). The 3$\alpha$-hydroxy group in the bile acid molecule is oxidized to a keto group by the enzyme reaction, while $\beta$-nicotinamide adenine dinucleotide (NAD) is reduced to nicotinamide adenine dinucleotide reduced from (NADH), which is subjected to voltammetric detection.

Figure 4-13 shows the cyclic semi-derivative and semi-integral voltammograms of NADH in phosphate buffer (pH 7.0) at a scan rate of 100 mV/sec with a glassy-carbon working electrode. It is clear that NADH is electroactive and irreversibly oxidized to NAD in this medium. For sensitive detection of NADH, the working electrode of the voltammetric detector should be set at the end potential, at which point the oxidation wave in the cyclic semi-derivative voltammogram is complete.

Figure 4-14 shows typical chromatograms of standard mixtures of free and conjugated bile acids. In these experiments, 30 mM $Na_4P_2O_7$ solution and 10 mM KNaHPO$_4$ solution containing 4 mM NAD-acetonitrile (48/30/22) and (40/30/30) were prepared as the mobile phases A and B, respectively. Thus, NAD was premixed in the mobile phase in order to increase

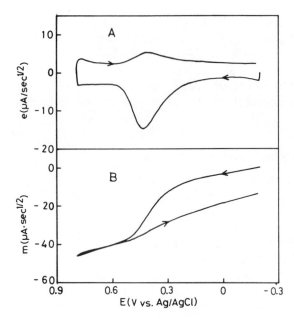

**FIGURE 4-13.** Electrochemical behavior of NADH in phosphate buffer (pH 7.0) on a glassy-carbon working electrode.[20] (A) Cyclic semi-derivative voltammogram, (B) cyclic semi-integral voltammogram. $e$ and $m$ represent the semi-derivative and semi-integral of the current with respect to time, respectively and $E$ is the electrode potential. (Reprinted from Goto et al.[20] with permission).

the sensitivity of detection. Mobile phase A was employed as the initial solution and stepwise changed to mobile phase B in 110 min. The working electrode was set at a potential of 0.80 V versus Ag-AgCl. The detection limits for free and conjugated bile acids were about 0.3–1 ng.

Hirata and co-workers[17] constructed a similar thin-layer cell by using a pressure-annealed pyrolytic graphite for the working electrode material, with a cell volume of 0.15 $\mu$l. The cell was applied to the determination of tyrosine and tryptophan metabolites in urine with a 60 m long microcapillary column packed with 30 $\mu$m porous silica particles. The measurements were conducted at a low flow rate of 1 $\mu$l/min using 0.2 M acetate buffer (pH 4.0) as the mobile phase, and the metabolites were detected by measuring the oxidation current with the potential of the working electrode set at 1.00 V versus a saturated calomel electrode.

For open-tubular HPLC Knecht and co-workers[19] developed a tubular type cell with a single graphite fiber electrode. The working electrode is constructed from a single carbon fiber with a 9 $\mu$m diameter and about a 0.7-mm length, which can be inserted with a micropositioner into the outlet end of the capillary column (15-$\mu$m ID × 265-cm length). A glass vessel is constructed to surround the fiber with 0.1 M KCl electrolyte solution and to furnish the silver-silver chloride electrode. The cell was tested in the oxidative mode by using a simple two-electrode arrangement. The electrolytic efficiency approached 100% at linear velocities below approximately 4 to 5 mm/sec, while detection limits for ascorbic acid, catechol, and 4-methyl-catechol were in the order of 1 fmol or $10^{-7}$M.

The structure of the tubular type cell built for microcolumns by Goto

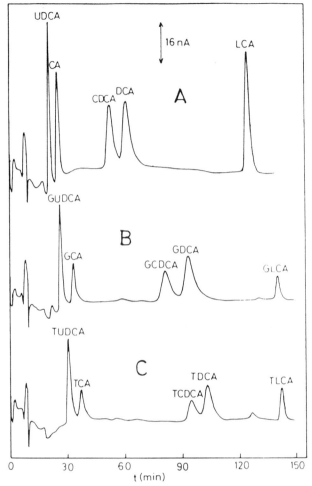

**FIGURE 4-14.** Typical chromatograms of a standard solution containing free and conjugated bile acids obtained by voltammetric detection with $3\alpha$-HSD post-column derivatization.[20] (A) Free bile acids, (B) glysine conjugates, (C) taurine conjugates. Applied potential: 0.80V versus Ag/AgCl. Mobile phase: see text. Flow rate: 8.3 $\mu l$/min. Peaks (each present in an amount of 80 ng): UDCA, ursodeoxycholic acid; CA, cholic acid; CDCA, chenodeoxycholic acid; DCA, deoxycholic acid; LCA, lithocholic acid; GUDCA, glycoursodeoxycholic acid; GCA, glycocholic acid; GCDCA, glycochenodeoxycholic acid; GDCA, glycodeoxycholic acid; GLCA, glycolithocholic acid; TUDCA, tauroursodeoxycholic acid; TCA, taurocholic acid; TCDCA, taurochenodeoxycholic acid; TDCA, taurodeoxycholic acid; TLCA, taurolithocholic acid.

and co-workers[18] is shown in Fig. 4-15. A single carbon fiber of 7 $\mu m$ diameter and 15 mm length is inserted into a fused-silica tube with an ID of 50 $\mu m$ to serve as the working electrode. The left-hand end of the fused-silica tube is connected with the outlet of the microcolumn and the solution is vented through a hole with a 1 mm diameter in the PTFE tube. A silver-silver chloride reference electrode and a platinum counterelectrode are

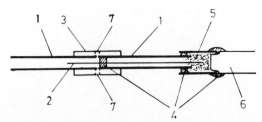

**FIGURE 4-15.** Structure of tubular type cell with a single carbon fiber electrode.[18] 1, Fused-silica tube (50 μm ID); 2, carbon fiber (7 μm diameter); 3, PTFE tube; 4, adhesive resin; 5, carbon paste; 6, electric wire; 7, hole. (Reprinted from Goto *et al.*[18] with permission.)

placed in the electrolyte solution drop near the outlet of the solution. This cell was utilized for the detection of catecholamines separated on a micro-column packed with silica-ODS of 3 μm particle size in a fused silica tube (0.35-mm ID × 50-mm length). Figure 4-16 shows a typical chromatogram of a standard solution. The detection limits for the four catecholamines were about 1–3 pg and the linear dynamic range was about $10^3$.

### 4.4.2. Miniaturized Dual Voltammetric Detectors

Miniaturized voltammetric detectors with two working electrodes have been reported only in the thin-layer type cell mode and can be classified into two types of configuration, serial and parallel, of the two working electrodes with respect to the flow axis.[21] In series configuration, the working electrodes are positioned along the flow stream on one side of the channel. In parallel configuration, the working electrodes are placed opposite one another on both sides of the channel.

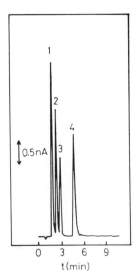

**FIGURE 4-16.** Typical chromatogram of a standard solution of four catecholamines obtained by voltammetric detection with tubular type cell.[18] Column: 0.35-mm ID × 50-mm long fused-silica tube packed with silica-ODS particles (3 μm). Mobile phase: methanol-phosphate buffer (pH 3.0) containing 0.4 mM 1-octane sulfonate and 0.2 mM EDTA (1:9) at 4.8 μl/min. Peaks (each 100 pg): 1, noradrenaline; 2, adrenaline; 3, dopa; 4, dopamine. (Reprinted from Goto *et al.*[18] with permission.)

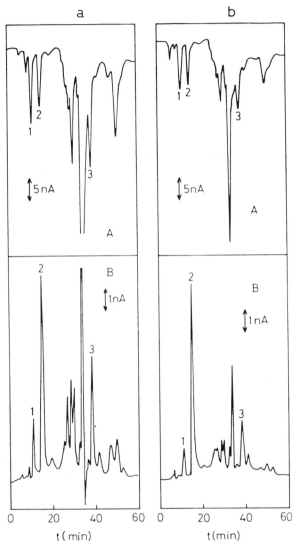

**FIGURE 4-17.** Selective detection of catecholamine metabolites in urine from two healthy individuals (a,b) with a series dual voltammetric detector.[24] (A) Anodic response and (B) cathodic response. Applied potentials (versus Ag/AgCl): anode, $+0.90$ V, cathode, $-0.05$V. Column: 0.5 mm ID $\times$ 154 mm long, containing silica-ODS particles (5 $\mu$m). Mobile phase: methanol–0.1 M phosphate buffer (pH 3.6) containing 0.1 mM EDTA (the methanol concentration was stepwise increased from 5% to 25% in 23 min). Flow rate: 5.6 $\mu$l/min. Peaks: 1, vanillylmandelic acid; 2, hydroquinone (int.std.); 3, homovanillic acid. (Reprinted from Goto *et al.*[24] with permission.)

The series dual voltammetric detector is analogous to the fluorescence detector, and the product of electrode reaction at the upstream working electrode is detected at the downstream working electrode. Goto and co-workers[22] built a sub-microliter thin-layer cell containing two glassy-carbon electrodes in series configuration. The cell was successfully utilized for the selective detection of catecholamines, indoleamine, and their metabolites in human urine, based on their electrochemical reversibility.[21-23] Figure 4-17 shows the chromatograms of catecholamine metabolites in urine from two healthy individuals, separated on a microcolumn packed with silica-ODS and using the series detector.[24] The urine sample was acidified, spiked with hydroquinone as the internal standard, and extracted with ethyl acetate. The organic solvent was evaporated to dryness with a stream of nitrogen gas, the residue being dissolved with the mobile phase. The extracted urine (0.3 $\mu$l) was injected as the chromatographic sample. Parts A and B in Fig. 4-17 are the anodic chromatograms obtained with the upstream working electrode and the cathodic chromatograms obtained with the downstream working electrode, respectively. The peaks appearing as the background of homo-vanillic acid are of particular interest in part A. By recording the re-reduction current, the interferences from the compounds responsible for these peaks could be removed, as shown in part B, on the basis of their electrochemical irreversibility. It is clear that vanillylmandelic acid and homovanillic acid in urine can be selectively detected simultaneously.

The parallel dual voltammetric detector is analogous to the photomultiplier tube, and the product of the electrode reaction at one working electrode can diffuse to the opposite working electrode where starting material may be created. Goto and co-workers built a thin-layer cell containing two glassy-carbon electrodes of 2 mm width and 1 cm length in parallel configuration.[25,26] For slower flow rates, catalytic amplification of detector response for the analytes, whose electrode reactions are reversible, may be achieved by recycling the redox couple between the two working electrodes. The current amplification in the cell was investigated at low flow rates by using ferricyanide as the analyte in Britton–Robinson buffer (pH 1.8). An amplification factor ranging from 2.4 to 19.5 was found when the flow rate changed from 11.2 to 1.4 $\mu$l/min.[25] The cell was successfully utilized for selective and sensitive detection of catecholamines in human serum separated on a 0.5 mmID × 150 mm long microcolumn packed with silica-ODS.[26] In this case, the upper and lower working electrodes were set at the potentials (versus Ag/AgCl) of +0.60 V and +0.20 V, respectively. The detection limit for catecholamines was about 3 pg, and only 200 $\mu$l of ultra-filtrated serum were required to determine catecholamines at normal level in human serum.[26]

## References

1. Snyder, L.R.; and Kirkland, J.J. "Introduction to Modern Liquid Chromatography," 2nd ed., Wiley-Interscience: New York, 1979, p. 126.

2. Ishii, D.; Asai, K.; Hibi, K.; Jonokuchi, T.; Nagaya, M. *J. Chromatogr.* **1977, 144,** 157.
3. Scott, R.P.W.; Kucera, P. *J. Chromatogr.* **1979, 169,** 51.
4. International Chromatography Guide, J. Chromatogr. Sci. **1986, 24,** 1G.
5. "1986 Buyers' Guide Edition", International Laboratory: **1986.**
6. Takeuchi, T.; Ishii, D.; Saito, M.; Hibi, K. *J. Chromatogr.* **1984, 295,** 323.
7. Yang, F.R. *J. Chromatogr.* **1982, 236,** 265.
8. Nagoshi, T. Faculty of Materials Science, Toyohashi University of Technology, personal communication, 1985.
9. Saito, M.; Wada, A.; Hibi, K.; Takahashi, M. *Industrial Research/Development,* **1983,** Apr. p 102.
10. Kucera, P.; Umagat, H. *J. Chromatogr.* **1983, 255,** 563.
11. Takeuchi, T.; Saito, S.; Ishii, D. *J. Chromatogr.* **1983, 258,** 125.
12. Schoeffel, D.M.; Sonnenschein, A.K. U.S. Patent 4 088 407, 1978.
13. Weinberger, R.; Sapp, E. *International Laboratory* 1984, Sep. p. 80.
14. Kucera, P. In "Microcolumn High-Performance Liquid Chromatography, Elsevier: Amsterdam, 1984, p. 70.
15. Hatano, H.; Yamamoto, Y.; Saito, M.; Mochida, E.; Watanabe, S. *J. Chromatogr.* **1973, 83,** 117.
16. Goto, M.; Koyanagi, Y; Ishii, D. *J. Chromatogr.* **1981, 208,** 261.
17. Hirata, Y.; Lin, P.T.; Novotny, M.V.; Wightman, R. M. *J. Chromatogr.* **1980, 181,** 287.
18. Goto, M.; Shimada, K.; Ishii, D. "Abstracts of Papers", 10th Discussion Meeting for HPLC, HPLC Discussion Committee, Nagoya, Japan, Jan 9, 1985, Abstr. 7.
19. Knecht, A.L.; Guthrie, E.J.; Jorgenson, J.W. *Anal. Chem.* **1984, 56,** 479.
20. Goto, M.; Kawaguchi, Y.; Ishii, D. "Abstracts of Papers", 44th Discussion Meeting of Analytical Chemistry, Japan Society for Analytical Chemistry, Nagasaki, Japan, Jun 6–7, 1983; Abstra. 1A12T.
21. Goto, M.; Sakurai, E.; Ishii, D. *J. Liq. Chromatogr.* **1983, 6,** 1907.
22. Goto, M.; Nakamura, T.; Ishii, D. *J. Chromatogr.* **1981, 226,** 33.
23. Goto, M.; Sakurai, E.; Ishii, D. *J. Chromatogr.* **1982, 238,** 357.
24. Goto, M.; Sakurai, E.; Ishii, D. "Abstracts of Papers", 32nd Annual meeting of the Japan Society for Analytical Chemistry, Niigata, Japan, Oct 3–7, 1983; Abstr. 3A05.
25. Goto, M.; Zou, G.; Ishii, D. *J. Chromatogr.* **1983, 268,** 157.
26. Goto, M.; Zou, G.; Ishii, D. *J. Chromatogr.* **1983, 275,** 271.

# Hyphenated Systems That Employ Microscale Columns

## K. Jinno and S. Tsuge

### 5.1. Microcolumn HPLC-IR Spectrometry

### 5.1.1. Introduction

Most compounds are able to absorb energy in the mid-IR region of the electromagnetic spectrum, and this fact can be utilized for compound identification. Since no commercially available LC detectors give structural information on the chromatographed compounds, IR spectroscopy can serve as an LC detector with great growth potential and both universal and chemically specific detection capabilities.

However, fundamental problems of LC-IR compatibility have been encountered, and the development of the proper interfacing has proved to be considerably difficult. In comparison with previously described absorption methods, IR absorption processes are relatively weaker and at least a few hundred nonograms of a sample are required to obtain acceptable IR spectra. In addition to this relatively poor sensitivity, many solvents used as mobile-phase components absorb strongly in the mid-IR region and in some spectral regions they are opaque. In order to maintain a sufficiently high transmittance, which will allow solute absorption bands to be observable at most wavelengths across the spectrum (approximately 40% at the frequency of interest), the pathlength of the flow cell must be kept short, typically around 100 $\mu$m or less. As the path length of the flow cell is increased in order to improve detectability, the detection limit may be conversely increased in the spectral windows because of the increase of absorp-

***K. Jinno*** School of Material Science, Toyohashi University of Technology, Toyohashi 440, Japan. ***S. Tsuge*** Department of Synthetic Chemistry, Faculty of Engineering, Nagoya University, Chikusa-ku, Nagoya 464, Japan.

tions by the organic solvents. Therefore, a compromise, which will give maximum cell thickness consistent with solvent transmission over all the bands of interest, is sought.

In spite of these disadvantages, the possibility of interfacing LC with IR has been investigated in great depth. Two fundamentally different types of LC-IR interfaces have been developed. In the first, the effluent from the column is passed directly through a flow cell, while the second approach involves the use of solvent elimination techniques. In both approaches, the high flow rate of the mobile phase required for conventional LC columns (for 4.6 mm ID columns, typically 1 ml/min) has represented a major limitation. When a flow cell is used, minor peaks elute at a concentration that is well below the detection limit for IR-absorption spectrometry. This is partly because of reduced transmittance in the spectral regions, where the mobile-phase components absorb, and partly because only a very small fraction of each eluate is present in the flow cell at any instant during the measurements. In solvent elimination techniques, mobile phase solvents must have higher volatility than those of the solutes chromatographed. (The generation of large volumes of solvent vapor also presents an environmental hazard unless great care is taken.)

Microcolumn LC permits the use of lower flow rates to achieve the same linear velocity and approximately the same chromatographic efficiency as with conventional 4.6 mm ID columns containing the same packing materials. Typical flow rates operated with microcolumns are between 1 and 100 $\mu$l/min. These low flow rates could allow on-line interfacings between LC and several spectrometric detection systems. The use of expensive solvents for the mobile phase becomes less prohibitive as the flow rates decrease. As a result of the lower flow rates, the increased concentration of minor components and the possibility of using deuterated solvents eliminates or reduces several of the problems associated with LC-IR interfaces that use flow cells. Microcolumn LC, and the corresponding lower flow rates, can also solve the problems associated with the solvent elimination approaches found in LC-IR interfaces, because solvents that are prohibited in IR can be eliminated easier.

In this section, the recent approaches in microcolumn LC-IR interfaces will be reviewed in order to strengthen the practical potential of microcolumn LC. Among IR spectrometers, Fourier-transform IR (FTIR) spectrometers have the features of very short scan times and signal-averaging, along with spectral subtraction capabilities. With these merits, FTIR, as opposed to conventional dispersive IR, is commonly considered the easier and more effective means of accomplishing LC-IR interfaces, although the cost of conventional IR versus FTIR still precludes the use of the latter in many laboratory situations. From the point of performance, FTIR is preferred, but there are no differences between LC-IR and LC-FTIR in the interfacing approaches. Therefore the following discussion is equally applicable for both conventional IR and FTIR.

### 5.1.2. Flow Cell Approach

Only the flow-cell technique meets the final requirements for chromatographic detection, in that it possesses the capability for real-time analysis. Unfortunately, typical LC solvents have strong absorption bands in the mid-IR region and this approach can be quite difficult. In the direct flow-cell approach, there are many questions to be considered before successful interfacing between LC and IR can be achieved. The first question concerns the selection of solvents to be used as LC mobile-phase components. Flow cells with smaller volume and longer path length are preferred for the purpose of IR monitoring. Conversely the path length must be decreased to reduce spectral interferences caused by the solvents when IR spectra of solutes are desired. In order to compensate for the sensitivity loss due to the smaller path length, the amount of a sample injected into an LC column must be increased, or more sensitive optical system and detector must be used. Since the former step results in the overloading of LC columns, while making improvement of the optical systems more difficult, the most transparent chromatographic solvents an optimum flow cell path length must be selected. These are very difficult problems to solve and they are the main reasons why the flow cell approach for conventional LC-IR interfacings is often considered to be difficult.

If our interest is limited to the detection of sample components having $C-H$ and/or $C=O$ fragments in their chemical structures (an almost universal feature of organic compounds) many solvents can be used without any interference, although they might otherwise be prohibitively costly, e.g., deuterated and/or perhalogenated solvents. Because solvent consumption is quite low in microcolumn LC, these costly but better quality IR solvents can be used for LC separations.[1-2]

The second question concerns the flow-cell material. With most organic solvents that are used as the mobile phase, the standard alkali salt window materials, such as NaCl, KBr, CsBr, and CsI, are optimal. ZnS, AgCl, and KRS-5 can be used with aqueous systems, but their higher refractive indices result in lower transmission and interference fringing. PTFE is also a good cell material, compatible with most chromatographic samples and solvents.[3] There are no definitive differences in the selection of cell materials used in conventional LC-IR or microcolumn LC-IR. However, microcolumn LC allows the use of smaller path-length cells and induces lower detection limits because of the increased concentration of the eluted component.

The third question important in LC-IR is where should a small-beam diameter be used at the focus. If the normal beam is used, no special alignment is necessary and LC-IR can conveniently be performed in the sample compartment. However, the sensitivity will not be enough for an LC detector. Condensing optics can be used to obtain a smaller beam size to improve or maintain the sensitivity. Generally, commercially available beam condensers are satisfactory for these purposes. The small-scale dimension of

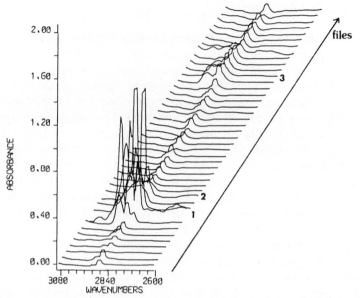

**FIGURE 5-1.** Normal-phase separation of steroids monitored by FTIR. Column: 0.5-mm ID × 110-mm length packed with JASCO FineSIL-5 (silica, 5 $\mu$m). Mobile phase: $d_{12}$-cyclohexane/$d_1$-chloroform/$d_4$-methanol (90.3/9.1/0.6 V/V). Flow rate: 8 $\mu$l/min. Detection: FTIR, accumulation 10 times (JEOL JIR-40X). Peaks: 1, 5$\alpha$-cholestane; 2, 5$\alpha$-cholestan-3-one; 3, cholesterol.

microcolumns permits use of such optical options more easily than in the case of conventional columns.

One can conclude from the above considerations that microcolumn LC is the most suitable way to combine LC with IR spectrometry via the flow cell approach. Some typical examples of microcolumn LC-IR measurements are demonstrated in the rest of this section.

For absorption or normal phase separations, saturated alkanes and cycloalkanes are quite common base solvents, into which small percentages of a polar solvent are added as a retention modifier. However, none of them has any reasonable transparency in the mid-IR regions. Some exceptions are represented by the fluorinated or chlorinated solvents, although they are limited by the coverable polarity range and their high costs. Figure 5-1 shows a three-dimensional FTIR chromatogram for a mixture of 5$\alpha$-cholestane, 5$\alpha$-cholestan-3-one, and cholesterol on silica with deuterocyclohexane/deuterochloroform/tetradeuteromethanol as the mobile phase. The effluent from the microcolumn could be identified by the characteristic IR-absorption bands observed in these three-dimensional spectra, in spite of the blacked-out spectral regions that were caused by the C−D and C−C absorptions.

The separation in normal phase mode was monitored at the C−H stretching region for a mixture of two alkanes, as shown in Fig. 5-2. A typical

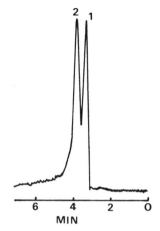

**FIGURE 5-2.** Normal-phase separation of alkanes monitored by IR. Column: 0.5-mm ID × 60-mm length packed with JASCO SS-05 (silica, 5 μm). Mobile phase: FC-78 (*N*-trifluoromethyl perfluoromorphorine; 3M Company. Flow rate: 16 μl/min. Detection: IR at 2940 cm$^{-1}$. Peaks: 1, *n*-hexane; 2, *n*-dodecane.

perfluorocarbon solvent, FC-78 was used as the mobile phase. If normal organic solvents had been used as the mobile phase, it would have been impossible to monitor the chromatogram at the C—H stretching absorption regions.

The next examples, shown in Figs. 5-3 and 5-4, are separations in reversed-phase mode. It is well known that the very polar solvents used for reversed-phase LC, such as water, methanol, and acetonitrile, absorb IR radiation much more strongly than any other organic solvent used in normal-phase mode. Although much of the fingerprint region of the spectrum was obscured, C—H and C=O stretching bands could be observed with the use of microcolumns coupled with the use of deuterated solvents.[1-3]

In the example shown in Fig. 5-3, the separation of three organic acids using $CD_3CN$ as the mobile phase was monitored at the C—H stretching

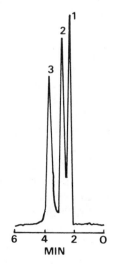

**FIGURE 5-3.** Reversed-phase separation of fatty acids monitored by IR. Column: 0.5-mm ID × 90-mm length packed with Chemcosorb ODS-H (7 μm; Chemco, Osaka, Japan). Mobile phase: $CD_3CN$. Flow rate: 16 μl/min. Detection: IR at 2937 cm$^{-1}$. Peaks: 1, caprylic acid; 2, capric acid; 3, lauric acid.

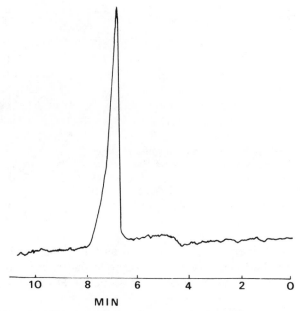

**FIGURE 5-4.** Reversed-phase separation of 1-propanol, monitored by IR. Column: 0.5-mm ID × 90-mm length packed with Chemco ODS-H (7 μm). Mobile phase: $D_2O$. Flow rate: 4 μl/min. Detection: IR at 3000 cm$^{-1}$.

region. Fig. 5-4 is another example of IR monitoring in the reversed-phase mode, where propyl alcohol was chromatographed using $D_2O$ as the mobile phase.

These examples show that IR monitoring in microcolumn LC can be performed using the direct flow cell approach coupled with the use of deuterated or perfluorinated solvents as the mobile phase. These solvents would be prohibitively costly in conventional LC-IR interfaces.

### 5.1.3. Solvent Elimination Approach

The intrinsic disadvantage present in LC-IR using the direct flow cell approach can be overcome by eliminating the solvent prior to the IR measurements. The most successful application of this approach was reported by Kuehl and Griffiths in 1979.[4] Their method involved a device in which the effluent from the LC column was concentrated and then dropped onto KCl powder. The remaining solvent rapidly evaporates leaving only the desired solute on the KCl and the diffuse reflectance spectrum of the solute can be measured. In 1980, Kuehl and Griffiths proposed a microcomputerized interface for this technique.[5]

Although solvent elimination approaches were thought to be a solution to all the problems associated with the flow cell-LC-IR combination, they still have a few limitations. The first is that the solutes have to be significantly less volatile than the mobile phase. The second limitation is that water cannot be successfully eliminated because of its high surface tension and latent heat of vaporization. The first problem, however, is not crucial in LC because volatile solutes are usually separated by GC. Therefore the most significant problem is the elimination of water.

The first microcolumn LC-IR results from experiments using solvent elimination techniques were reported by Jinno *et al.*[6,7] and Fujimoto *et al.*[8] in 1982. Column effluent was deposited as a "buffer-memory" on the KBr crystal plate. After the chromatogram was complete, the plate was automatically and simply transferred to the IR spectrometer and the transmittance spectra were continuously measured across that region of the plate where the peaks had been deposited. Detection limit was usually found to be between 100 ng and 1 $\mu$g, and no spectral region was obscured by the solvent.

The schematic of the interface device used is shown in Fig. 5-5. The effluent from the microcolumn was deposited on a KBr crystal plate that was slowly moved across the exit of a stainless-steel connecting tube. The solvent was evaporated with help of a warm nitrogen gas flow, and the collection speed was controlled according to the LC conditions used. The buffer-memory crystal was then placed into the IR beam. Figure 5-6 is the photograph of the KBr crystal plate and Fig. 5-7 shows the layout of the interface.

Because of the higher volatility of the organic solvents, which are commonly used as the mobile phase in the normal-phase and size-exclusion modes, this approach has been found successful in these modes. To demonstrate the buffer-memory technique, a mixture of three polystyrene standards was eluted through a microcolumn. The resulting IR chromatogram, monitored at 698 cm$^{-1}$, is shown in Fig. 5-8. It can be seen that the IR chromatogram obtained by the buffer-memory technique is quite good and shows no interferences from the solvent.

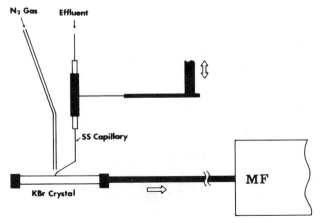

**FIGURE 5-5.** Interface for microcolumn LC-IR using the solvent elimination technique. MF, microfeeder (MF-2; Azuma Electric, Tokyo, Japan).

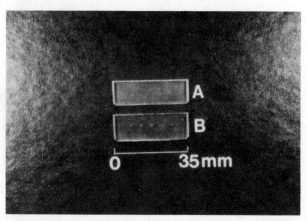

**FIGURE 5-6.** KBr buffer-memory plate. (A) Deposited di-*n*-propylketone-2, 4-dinitro-phenylhydrazone. From the left to the right: 1, 3, 4, and 6 μg. (B) Deposited di-*n*-propylketone-2, 4-dinitro-phenylhydrazone. 2 μg × 5 times.

The problem related to the elimination of water is serious if alkali salts are used as the collecting medium. However, measurements of LC-IR in reversed-phase separations are possible when a stainless-steel wire net is used as the collecting medium instead of a KBr crystal.

Figure 5-9 shows the three-dimensional FTIR chromatogram representing the separation of caffeine, aspirin, and phenacetin. No interference is observed from absorption of the mobile phase components, methanol and water. This example strongly indicates that the buffer-memory technique can be a universal method for microcolumn LC-IR interfacing in both normal- and reversed-phase separations.

In summary, IR monitoring is useful for the detection of solutes separated by LC, particularly as good selectivity can be gained by using specific

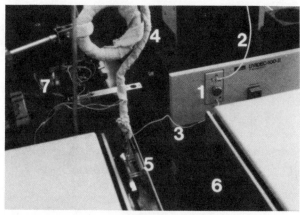

**FIGURE 5-7.** Layout of the buffer-memory interface. 1, UV detector; 2, column; 3, connecting tube; 4, heater for warm nitrogen gas; 5, KBr crystal plate; 6, IR spectrometer; 7, microfeeder.

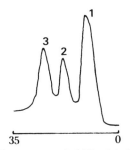

Distance moved of KBr plate (mm)

**FIGURE 5-8.** Size-exclusion separation of polystyrene standards monitored by IR via the KBr buffer-memory technique. Column: 1-mm ID × 220-mm length packed with Toyo Soda TSK Gel-G3000H (Tokyo, Japan). Mobile phase: THF. Flow rate: 8 μl/min. Detection: IR at 698 cm⁻¹. Peaks: 1, 37,000 mol. wt.; 2, 10,200 mol. wt.; 3, 2,800 mol. wt.

wavelengths. The simpler, and certainly less expensive, interfacing systems that have effective use of the features of microcolumn LC are satisfactory for most applications in spite of their relatively poor sensitivity.

## 5.2. Microcolumn HPLC-Mass Spectrometry

### 5.2.1. Introduction

Combined HPLC-MS has been investigated in detail.[9-18] Although some types of HPLC-MS systems are now commercially available and many promising applications have been reported, we have to recognize that HPLC/MS is not yet sufficiently matured so that the majority of the potential users could utilize it. This state of the art reflects the fact that the problems involved in establishing a really practical HPLC-MS system are far more difficult than those of GC-MS systems. One problem is that the thermally labile and/or involatile sample molecules are continuously and stably evaporated from the HPLC effluent and then softly ionized for the subsequent MS separation. A second difficulty is that conventional mass spectrometers cannot accept the excess volume of the vaporized components including as diverse as even buffers when conventional HPLC columns with a flow rate of about 1 ml/min are utilized. The use of packed microbore, open-tubular, or packed capillary columns instead of the conventional columns, however, has greatly diminished this difficulty, since the reduced flow rates enable the total effluent from the HPLC column to be more easily introduced to the mass spectrometer, often with only minor modification of the associated system.

At present, the two most popular HPLC-MS methods are the mechanical transfer method (by a moving belt) and the direct introduction method. The latter method includes such varieties as introduction either through a diaphragm with a pinhole or through an extended capillary, vacuum nebulization with enrichment, and thermospray.

**FIGURE 5-9.** Reversed-phase separation of caffeine, aspirin, and phenacetin, monitored by FTIR via the buffer-memory technique. Column: 0.5-mm ID × 250-mm length packed with Nomura Chemicals Develosil ODS-10 (10 μm; Seto, Japan). Mobile phase: methanol-water (60/40). Flow rate: 4 μl/min. Detection: FTIR, 100 times accumulation (JEOL JIR-40X). Peaks: 1, caffeine; 2, aspirin; 3, phenacetin.

This section will discuss the basic principles of the typical HPLC-MS methods and their recent applications, focusing on the use of micro-HPLC.

### 5.2.2. Mechanical Transfer Method

Figure 5-10 illustrates a typical moving belt HPLC-MS interface supplied by Finigan.[19] The effluent from the HPLC column is continuously deposited on a moving Kapt® (polyimide) belt with a 3 mm width. After solvent evaporation, the belt is transferred through vacuum locks at a constant speed (2–3 cm/sec) into the ionization chamber, where the solute components on the belt are rapidly evaporated by a flash vaporizer. A clean-up heater is provided to remove the solute residue. Although both EI and CI modes can be used for the ionization of the vaporized components, serious problems are sometimes encountered due to the thermal decomposition of the thermally labile compounds on the belt. The solvent capacity of the belt changes depending on the nature of the solvents used for HPLC. Occasionally, effluent splitting permits higher mobile-phase flows. For a mobile phase containing higher water contents, the capacity sometimes falls to as low as 0.05 ml/min. Spray deposition is often effective in such cases.[20]

Figure 5-11 shows a typical mass chromatogram obtained by the HPLC-MS system interfaced by a moving belt, where antibiotics such as spectinomycin and actinamine were separated by a reversed-phase column with gradient elution and detected by a methane chemical-ionization mass spectrometer.[19]

Applications of the mechanical transfer HPLC-MS system combined with continuous liquid-liquid extraction include various less volatile compounds such as fatty acids, alcohols, amines, and pesticides.[21]

Another current trend in HPLC-MS interfacing is to combine the moving-belt method with surface ionization techniques, such as secondary ion mass spectrometry (SIMS),[22,23] fast atom bombardment (FAB),[24] and laser desorption.[25] By these methods, the initial limitations of the belt interface for thermally-labile compounds might be drastically alleviated.

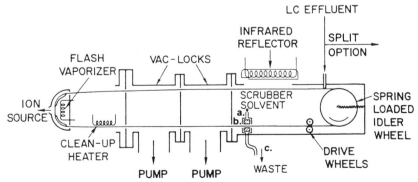

**FIGURE 5-10.** Schematic diagram of a moving belt interface. (a) Scrubbing solution enters; (b) extraction of salt deposits; (c) waste. (Courtesy of Finigan Corporation.)

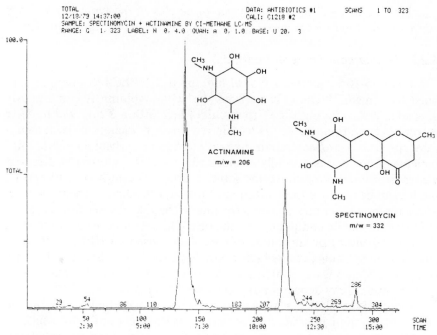

**FIGURE 5-11.**   Mass chromatogram for ions specific to actinamine and spectinomycin.[19]

### 5.2.3. Introduction through a Diaphragm with a Pinhole or through an Extended Capillary

The system that introduced utilization of a diaphragm with a pinhole was originally developed for the connection of conventional HPLC columns with the mass spectrometer. A portion (approximately 1%) of the column effluent was fed into the ion source of the mass spectrometer, where the solvent vapor acted as the reagent gas for chemical ionization. However, Henion et al.[26] and Eckers et al.[27] reported a micro-HPLC-MS probe (shown in Fig. 5-12) by which the total effluent (10–60 μl/min) from a micro-HPLC column was introduced through a diaphragm with a 5-μm pinhole into the chemical ionization source. Volatile buffers such as ammonium hydroxide, trimethylamine, ammonium acetate, and trifluoroacetic acid can safely be utilized in this system. Various steroids were successfully analyzed by this system.

Bruins et al.[28] developed another interface for micro-HPLC-MS coupling, which consisted of an extended capillary. Figure 5-13 shows the schematic diagram of the interface. The total effluent (approximately 10 μl/min) from the micro-HPLC column was directly introduced into the chemical ionization source through a fused-silica capillary (50 μm ID × 70 cm length) with a copper block serving as the heat conductor from the wall of the ion-

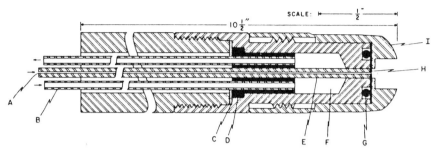

**FIGURE 5-12.** Schematic diagram of a direct introduction interface with pinholed diaphragm.[26] (A) Effluent from the micro-HPLC; (B) water cooling inlet; (C) PTFE seal; (D) throughput tube collect; (E) 0.004-inch ID × 0.062-inch OD stainless-steel tube; (F) water cooling chamber; (G) O-ring; (H) diaphragm with 5-$\mu$m pinhole; (I) removable endcap.

ization chamber. Figure 5-14 shows the UV chromatogram and the mass chromatogram of four similar drug components separated on a reversed-phase column at a flow rate of 8 $\mu$l/min.

Both the diaphragm and capillary column introduction sometimes encounter a problem when the direction of the jet deviates from a straight line when the interface is exposed under reduced pressure.[28] This unstable phenomenon is closely related to the irregular shape of the hole and is more frequently observed for the diaphragm interface with a small pinhole (1–10 $\mu$m), since the irregularity of the hole can cause extensive abnormal nebulization with a very thin diaphragm pinhole under vacuum conditions. Therefore, in order to achieve a long-term stable operation, the "optimum" state of the orifice and the capillary has to be strictly controlled.

### 5.2.4. Introduction through Vacuum Nebulization with Enrichment

A schematic diagram of the nebulizing interface is shown in Fig. 5-15 together with the magnified portion of the nebulizing tip.[29] The housing case of the cooling water jacket was made of a low heat-conducting glass-ceramic (Macor®). In the most recent design, as shown in Fig. 5-16, the whole effluent (10–50 $\mu$l/min) from the micro-HPLC column is conducted to the

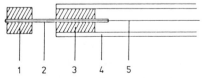

**FIGURE 5-13.** Schematic diagram of direct introduction interface with extended capillary.[28] 1, copper block (4.9-mm OD); 2, stainless-steel tube (0.25-mm ID × 0.5-mm OD); 3, PTFE insulator; 4, stainless-steel tube (4.6-mm ID × 6.4-mm OD); 5, fused-silica capillary (0.050-$\mu$m ID × 70-cm long).

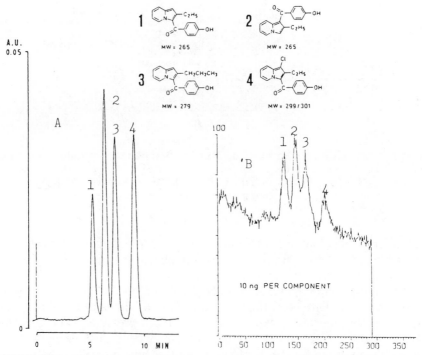

**FIGURE 5-14.** Chromatograms of four drug components.[28] (A) Chromatogram obtained by UV detection at 390 nm. (B) Mass chromatogram obtained by total ion current monitoring (m/z = 150 − 350). Mobile phase: acetonitrile/water (70/30), at 8 μl/min.

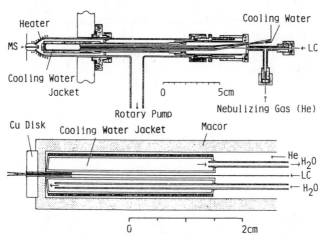

**FIGURE 5-15.** Schematic diagram of the vacuum nebulizing interface.[29]

top of the nebulizing tip (12 $\mu$m ID fused-silica capillary) and finely nebulized by a jet stream of the nebulizing gas, which was supplied through a gap
between the tip and a stainless steel sheath (0.63-mm ID $\times$ 0.33-mm OD).
After most of the solvent vapor was evacuated through a rotary pump, the
enriched components were introduced continuously into the chemical ionization source of a quadrupole mass spectrometer. When the nebulizing gas
was used, the effluent was more finely nebulized and the resulting mass fragmentogram of the solute became very stable (Fig. 5-16). The long-term stability was also drastically improved by this enforced nebulization. Using
this HPLC-MS system, applications were extended to various less volatile
and/or thermally-labile compounds such as amino acids, peptides,
saccharides,[29] pesticides, steroids,[30] and free fatty acids.[31]

Figure 5-17 shows typical chromatograms of free fatty acids from bean
oil, separated by a micro-packed column at a flow rate of 16 $\mu$l/min and
detected by UV, at 210 nm, and by multi-ion detection (MID), with the
quadrupole mass spectrometer in the chemical-ionization mode. In this
application, the overlapping peaks of oleic acid and palmitic acid are dis-
criminately detected by MID on the mass chromatograms.

To obtain very long-term stability for interfaces assisted by self-spout-
ing and nebulizing, complete dust-free effluent introduction to the nebulizing nozzle is the most important factor, along with incorporating appropriate filters. In addition, the control of the temperature profile around the
nebulizing tip is also important.

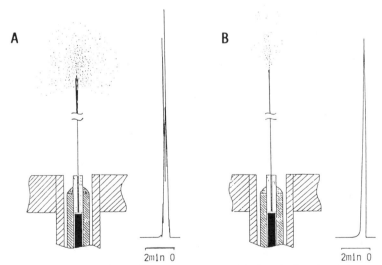

**FIGURE 5-16.** Effect of nebulizing gas on sample introduction.[17] (A) Without nebulizing gas
(He). (B) With nebulizing gas (He). Sample: aminopyrine. Mass chromatogram obtained by
selected ion monitoring.

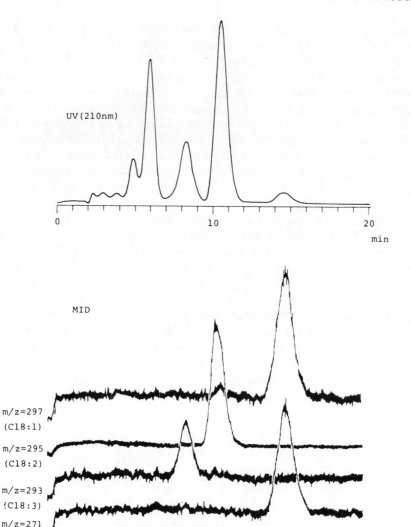

**FIGURE 5-17.** Chromatograms of free fatty acids from bean oil by UV-detection and multi-ion detection (MID).[31] Column: 0.5-mm ID × 145-mm length packed with ODS material (10 μm). Mobile phase: Methanol/water (9/1). Flow rate: 16 μl/min. UV detection at 210 nm.

## 5.2.5. Thermospray

Thermospray is defined as the complete or partial vaporization of a liquid stream, sometimes accompanied by ionization by heating as it sprays out from a capillary nozzle. Thermospray interfaces were first heated by laser, then by oxy-hydrogen torches, and most recently by electric cartridge heaters. Figure 5-18 illustrates a recent design of the thermospray interface.[32] The effluent from the HPLC column (up to a flow rate of 2 ml/min of an aqueous mobile phase) is thermosprayed directly into the ion source and the excess vapor is pumped away through a port opposite the spray nozzle (0.15-mm ID × 1.5-mm OD stainless-steel capillary). A portion of the resulting ions is introduced into a quadrupole mass spectrometer through an ion-exit aperture. When used with a mobile phase containing a significant concentration of ions in the solution (approximately $10^{-4}$ to 1 M), no external ionization is required.

Figure 5-19 shows an application of thermospray to peptide analysis. An arginine containing hexapeptide and an orinithine containing peptide were separated on a short ODS 3 $\mu$m column. Both peptides were monitored as doubly protonated ions.

Direct analysis of glucuronides[33] and pesticides[34] were also reported. Although this technique has not yet fully matured, it is becoming recognized as a promising practical approach for on-line HPLC-MS. The thermospray interfaces have so far been used entirely on quadrupole mass spectrometers. However, a recent paper[35] reports the direct coupling of this technique with a magnetic mass spectrometer, the ion source of which is operated at high voltages (up to 4 KV). In this approach, the HPLC effluent is introduced to the thermospray vaporizer by using a 120 cm long, 150 $\mu$m ID fused-silica capillary to insulate the HPLC instrument from the high voltage. Such a system could extend the potential ability of HPLC-MS>

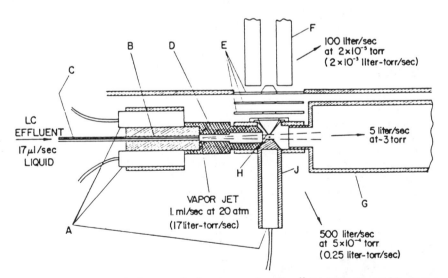

**FIGURE 5-18.** Schematic diagram of a thermospray interface.[32] (A) 100-watt cartridge heater; (B) copper block brazed to stainless-steel capillary; (C) 0.15 mm ID × 1.5 mm OD stainless-steel capillary; (D) thick-wall copper tube; (E) ion lenses; (F) quadrupole mass filter; (G) pump-out line to the mechanical pump; (H) ion-exit aperture; (J) source heater.

REL. INT.

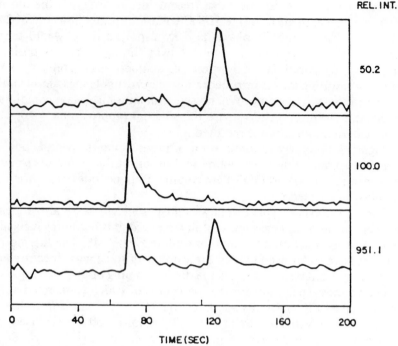

50.2

100.0

951.1

TIME (SEC)

**FIGURE 5-19.**   Mass chromatograms of peptides.[33] Column: 4.1-mm ID × 50-mm length packed with ODS material (3 μm). Mobile phase: acetonitrile/water containing 0.1M ammonium acetate. Gradient elution from 30 to 50% acetonitrile. (A) Orinithine peptide (Thr-Thr-vin-Gln-Orn-Try-NH₂; mol. wt., 738; m/z, 368–372). (B) Arginine peptide (Thr-Thr-Arg-Gln-Arg-Tyr-NH₂; mol. wt., 0.822; m/z, 409–413). (C) RIC (reconstructed ion chromatogram)

## References

1. Jinno, K.; Fujimoto, C.; Uematsu, G. *Amer. Labo.* **1984, 16(2),** 39.
2. Fujimoto, C.; Uematsu, G.; Jinno, K. *Chromatographia* **1985, 20,** 112.
3. Fujimoto, C.; Jinno, K. *J. High Resolut. Chromatogr./Chromatogr. Commun.* **1983, 6,** 374.
4. Kuehl, D.; Griffiths, P.R. *J. Chromatogr. Sci.* **1979, 17,** 471.
5. Kuehl, D.T.; Griffiths, P.R. *Anal. Chem.* **1980, 52,** 1394.
6. Jinno, K.; Fujimoto, C.; Hirata, Y. *Appl. Spectrosc.* **1982, 36,** 67.
7. Jinno, K.; Fujimoto, C.; Ishii, D. *J. Chromatogr.* **1982, 230,** 625.
8. Fujimoto, C.; Jinno, K.; Hirata, Y. *J. Chromatogr.* **1982, 258,** 81.
9. McFadden, W.H. *J. Chromatogr. Sci.* **1980, 18,** 97.
10. Tsuge, S. *Kagaku no Ryoiki Zokan* **1981, 132,** 132–170.
11. Curry, Z.F. *J. Liq. Chromatogr.* **1982, 5,** 257.
12. Arpino, P.J. *Trends in Anal. Chem.* **1982, 1,** 154.
13. McFadden, W.H. *Anal. Proc.* **1982, 19,** 258.
14. Games, D.E. *Anal. Proc.* **1982, 20,** 352.
15. Desiderio, D.M.; Fridland, G.H.; Stout, C.B. *J. Liq. Chromatogr.* **1984, 7,** 317.
16. Games, D.E.; Alcock, N.J.; McDonall, M.A.A. *Anal. Proc.* **1984, 21,** 24.
17. Tsuge, S. in "Microcolumn Separations", Novotny, M.V., Ishii, D., Eds.; **1985,** pp. 217–241.

18. Henion, J. in "Microcolumn Separation," Novotny, M.V., Ishii, D., Eds. **1985**, pp. 243–274.

19. Kelly, P.E. *Finigan Application Report* **no. AR8006,** 1980.

20. Hayes, M.J.; Lankmayer, E.P.; Voutos, P.; Karger, B.L.; McGuire, J.M. *Anal. Chem.* **1983, 55,** 1745.

21. Karger, B.L.; Kirby, D.P.; Vouros, P.; Foltz, R.L.; Hidy, B. *Anal. Chem.* **1979, 51,** 2324.

22. Benninghoven, A.; Eicke, A.; Junack, M.; Sichtermann, W.; Kirzek, J.; Peters, H. *Org. Mass Spectrom.* **1980, 15,** 457.

23. Smith, H.D.; Burger, J.E.; Johnson, A.L. *Anal. Chem.* **1981, 53,** 1603.

24. Hunt, D.F.; Bone, W.M.; Shabanowitz, J.; Rhodes, J.; Ballard, J.M. *Anal. Chem.* **1981, 53,** 1704.

25. Hardin, E.D.; Vestal, M.L. *Anal. Chem.* **1981, 53,** 1492.

26. Henion, J.; Wachs, T. *Anal. Chem.* **1981, 53,** 1936.

27. Eckers, C.E.; Skrabalak, O.S.; Henion, J.D. *Clin. Chem.* **1982, 28,** 1882.

28. Bruins, A.P.; Drenth, B.F.H. *J. Chromatogr.* **1983, 271,** 71.

29. Yoshida, Y., Yoshida, H.; Tsuge, S.; Takeuchi, T.; Mochizuki K. *J. High Resolut. Chromatogr./Chromatogr Commun.* **1980, 3,** 16.

30. Yoshida, H.; Matsumoto, K.; Itoh, K.; Tsuge, S.; Hirata, Y.; Mochizuki, K.; Kokubun, N.; Yoshida, Y. *Frezenius Z. Anal. Chem.* **1982, 311,** 674.

31. Matsumoto, K.; Yoshida, H.; Ohta, K.; Tsuge, S. *Org. Mass Spectrom.* **1985, 20,** 777.

32. Blakley, C.R.; Vestal, M.L. *Anal. Chem.* **1983, 55,** 750.

33. Pilosof, D.; Kim, H.Y.; Dyckes, D.F.; Vestal, M.L. *Anal. Chem.* **1984, 56,** 1236.

34. Voyksner, R.D.; Bursey, J.T.; Pellizzar, E.D. *Anal. Chem.* **1984, 56,** 1507.

35. Vestal, M.L. *Anal. Chem.* **1984, 56,** 2590.

# 6

# *Post-Column Derivatization in Microscale HPLC*

## *M. Senda and S. Higashidate*

### *6.1. Introduction*

In HPLC analysis, pre- and post-column derivatizations are commonly used to detect compounds that have no UV-VIS absorption or fluorescence emission and to improve selectivity and sensitivity. In pre-column derivatization, the derivatizing reaction is carried out before sample injection. Pre-column derivatization can be regarded as one of sample pretreatment methods, and there is no difference between conventional HPLC and micro-HPLC.

In post-column derivatization, column effluent containing the separated components is mixed with a derivatizing reagent. The effluent is then introduced into a reactor where the derivatizing reaction takes place. The reaction products are finally detected by a photometric, fluorometric, or electrochemical detector. Three types of reactors are used for this reaction: open-tubular, packed-bed tubular, and segmented-stream tubular. These reactors inevitably cause extracolumn peak broadening because sample solutes move slowly in them.

Therefore, in post-column derivatization in microscale HPLC, the extracolumn peak broadening caused by the reactor is a difficult problem to overcome. Deelder and co-workers[1-3] have intensively investigated and compared extracolumn peak broadening caused by those different types of reactors for conventional HPLC. In this chapter, their results will be applied to extracolumn peak broadening in microscale HPLC, and two common types of reactors, open-tubular and packed-bed, will be discussed. In this discussion, the reactors for a semi-microcolumn of about one-tenth the vol-

*M. Senda and S. Higashidate* Japan Spectroscopic Company, Ltd., Hachioji City, Tokyo 192, Japan.

ume of a conventional column will be described. However, reactors for a microcolumn of one-hundredth or less the volume of a conventional column will not be described here. The design of reactors, which meet extremely small peak volumes from such microcolumns, is very difficult.

## 6.2. Peak Broadening Caused by the Reactors

### 6.2.1. Open-Tubular Reactors

The time variance of the observed peak ($\sigma_{t(ob)}^2$) is expressed as a summation of the time variance of the peak by column contribution ($\sigma_{tc}^2$) and the time variance of the peak by open-tubular reactor contribution ($\sigma_{tr}^2$).

$$\sigma_{t(ob)}^2 = \sigma_{tc}^2 + \sigma_{tr}^2 \tag{1}$$

$\sigma_{tr}^2$ is given by

$$\sigma_{tr}^2 = \kappa \frac{d_t^2 \, t}{96 D_M} \tag{2}$$

where $d_t$ is the internal diameter of the reactor, $D_M$ is the molecular diffusion coefficient of the solute in the reactor fluid, and $t$ is its mean residence time in the reactor. $\kappa$ is related to the coiling of the reactor and is expressed by the Dean number ($Dn$) and the Schmit number ($Sc$). For $12.5 < (Dn \times Sc^{0.5}) < 250$, $\kappa$ is represented as[2,3]

$$\kappa = 5.6(Dn \times Sc^{0.5})^{-0.67} \tag{3}$$

These two numbers are given by

$$Dn = \frac{4F\rho}{\pi \eta d_t} (d_t/d_c)^{0.5} \tag{4}$$

$$Sc = \frac{\eta}{\rho D_M} \tag{5}$$

where $F$ is the total flow rate of the mobile phase and the reagent solution, $\rho$ is the density of the reactor fluid, $\eta$ is the viscosity of the fluid, and $d_c$ is the coil diameter. The pressure drop ($\Delta P$) is approximated as

$$\Delta P = \frac{512 \, \eta \, F^2 t}{\pi^2 d_t^6} \tag{6}$$

and the mean residence time ($t$) is expressed as

$$t = \frac{\pi d_t^2 L}{4F} \tag{7}$$

where $L$ is the reactor coil length. Using equations 2, 3, 4, 5, 6, and 7 for a $d_t$ of 0.1 mm and an $F$ equaling $3.333 \times 10^{-3}$ cm$^3$/sec, which are the respective typical reactor internal diameter and flow rate values in semi-micro HPLC. Assuming that $D_M$ equals $10^{-5}$ cm$^2$/sec, $\rho$ equals 1 g/cm$^3$, $\eta$ equals $10^{-2}$ g/cm $\times$ sec$^2$, and $d_c$ equals 2 cm, we obtain

$$t = 0.02355L \qquad (8)$$
$$\sigma_{tr}^2 = 0.02762t \qquad (9)$$
$$\Delta P = 5.770 \times 10^6 t \qquad (10)$$

For a $d_t$ of 0.25 mm, we obtain

$$t = 0.1472L \qquad (11)$$
$$\sigma_{tr}^2 = 0.2346t \qquad (12)$$
$$\Delta P = 0.02363 \times 10^6 t \qquad (13)$$

If we use the semi-microcolumn described in Chapter 2 Table 2-1, which has a dimension of 1.5-mm ID $\times$ 250-mm length and a theoretical plate number of 10,000, for a mobile phase flow rate of 0.1 ml/min (1.667 $\times$ 10$^{-3}$ cm$^3$/sec), a peak with the capacity factor of $k' = 1$ has a volume of 24 $\mu$l, and $\sigma_{tc}$ of this peak is 3.72 sec. If the peak broadening of this peak caused by the reactor is allowed to be smaller than 10%, that is, $\sigma_{t(ob)}/\sigma_{tc} < 1.1$, we can calculate the maximum time variance contributed by the reactor, $\sigma_{tr(max)}^2$, using equation 1.

$$\sigma_{tr(max)}^2 = 2.91 \text{ sec}^2 \qquad (14)$$

If the pressure drop caused by the reactor ($\Delta P$) is allowed to be smaller than 100 bar (100 $\times$ 10$^6$ g/cm $\times$ sec$^2$) for a reagent solution flow rate of 1.667 $\times$ 10$^{-3}$ cm$^3$/sec (total flow rate = 3.333 $\times$ 10$^{-3}$ cm$^3$/sec), the maximum reaction times ($t_{max}$) and the maximum reactor lengths ($L_{max}$) can be calculated for reactors with 0.1mm or 0.25-mm ID (as shown in Table 6-1) using equations 8, 9, 10, 11, 12, 13, and 14.

As shown in Fig. 6-1, the relationship between the reaction time, the reactor tube length, and the peak broadening caused by the reactor under the conditions of $\sigma_{t(ob)}/\sigma_{tc} < 1.1$ and $\Delta P < 100$ bar can be obtained using equations 1, 8, 9, 10, 11, 12, 13, and 14.

**TABLE 6-1**

Maximum Reaction Times and Maximum Reactor Lengths for Open-tubular Reactors[a]

|  | Internal diameter of the reactor (mm) | |
| --- | --- | --- |
|  | 0.1 | 0.25 |
| Maximum reaction time ($t_{max}$; sec) | 17.3 | 12.4 |
| Maximum reactor length ($L_{max}$; cm) | 740 | 80 |

[a]10 % increase in peak broadening and 100 bar pressure drop are allowed.

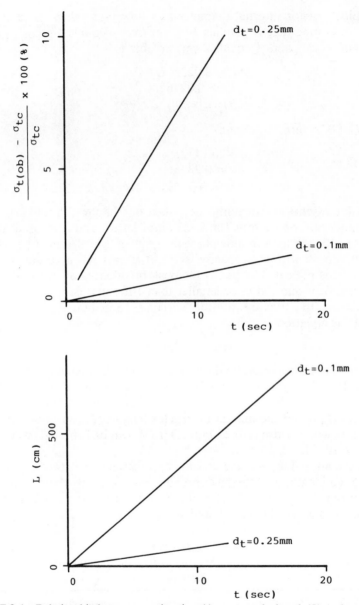

**FIGURE 6-1.** Relationship between reaction time ($t$), reactor tube length ($L$), and peak broadening. $\sigma_{t(ob)}$, standard deviation of the observed peak in time units; $\sigma_{tc}$, standard deviation of the peak by column contribution; $d_t$, internal diameter of the reactor tube.

### 6.2.2. Packed-Bed Reactors

Packed-bed reactors usually consist of a stainless-steel tube packed with nonporous spherical glass beads. The reduced plate height of the packed-bed reactor ($h$) is expressed as

$$h = \frac{2\gamma}{\nu} + A\nu^{0.33} \qquad (15)$$

where $\nu$ is the reduced velocity of the reactor fluid, $\gamma$ is the tortuosity factor of the reactor, and A is a constant that depends on the reactor bed geometry. In a good packed-bed reactor, $\gamma$ and A are 0.8.[3] $\nu$ is defined as

$$\nu = \frac{u d_p}{D_M} \qquad (16)$$

where $d_p$ is the particle diameter, $D_M$ is the molecular diffusion coefficient of the solute in the reactor fluid, and $u$ is the linear velocity of the reactor fluid. $u$ is expressed as

$$u = \frac{L}{t} \qquad (17)$$

where $L$ is the reactor column length and $t$ is the mean residence time of the solute in the reactor. The theoretical plate number ($N$) is expressed as

$$N = \frac{L}{H} = \frac{L}{h d_p} \left(\frac{t}{\sigma_{tr}}\right)^2 \qquad (18)$$

where $H$ is the plate height of the reactor column and $\sigma_{tr}^2$ is the time variance of the peak by reactor contribution. Using equations 15, 16, 17, and 18, we obtain

$$\sigma_{tr}^2 = \frac{2\gamma D_M t^3}{L^2} + \frac{A t^{1.66} d_P^{1.33}}{L^{0.67} D_M^{0.33}} \qquad (19)$$

**TABLE 6-2**
Maximum Reaction Times and Maximum Reactor Lengths for Packed-bed Reactors[a]

|  | Particle size ($\mu$m) | | | | |
|---|---|---|---|---|---|
|  | 10 | 20 | 30 | 40 | 50 |
| Maximum reaction time ($t_{max}$; sec) | 129 | 168 | 98 | 67 | 50 |
| Maximum reactor length ($L_{max}$; cm) | 51 | 66 | 39 | 26 | 20 |

[a]10 % increase in peak broadening and 100 bar pressure drop are allowed. Internal diameter of the reactor: 1.5 mm.

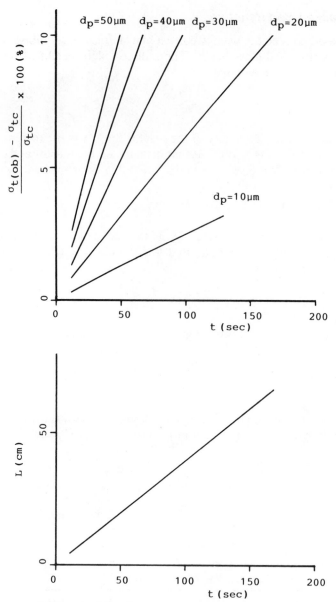

**FIGURE 6-2.** Relationship between reaction time ($t$), reactor column length ($L$), and peak broadening. $\sigma_{t(ob)}$, standard deviation of the observed peak in time units; $\sigma_{tc}$, standard deviation of the peak by column contribution; $d_p$, particle diameter.

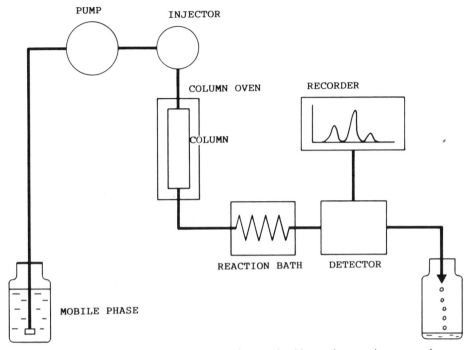

**FIGURE 6-3.** Flow diagram of the adenine nucleotide analyzer using post-column derivatization.

The pressure drop caused by the reactor ($\Delta P$) is expressed as

$$\Delta P = \frac{\eta L^2}{k_0 t d_P^2} \qquad (20)$$

where $k_0$ is the permeability constant of the type of packing used.

The mean residence time of the solute in the reactor ($t$) is expressed as

$$t = \frac{\epsilon \pi d_c^2 L}{4F} \qquad (21)$$

where $d_c$ is the internal reactor column diameter, $F$ is the total flow rate of the mobile phase and the reagent solution, and $\epsilon$ is the reactor column poros-

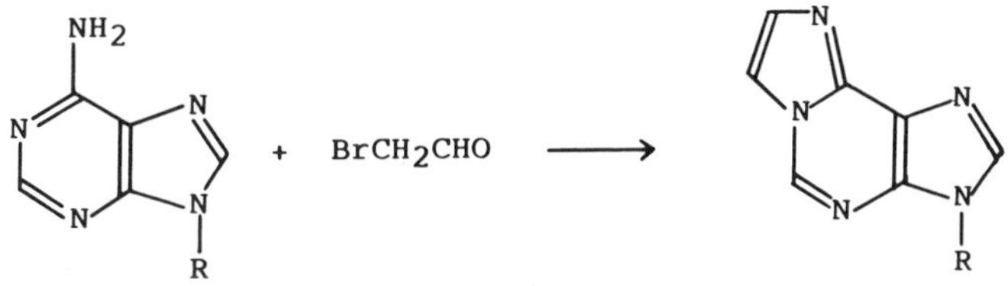

**FIGURE 6-4.** Reaction of bromoacetaldehyde with adenine nucleotides.

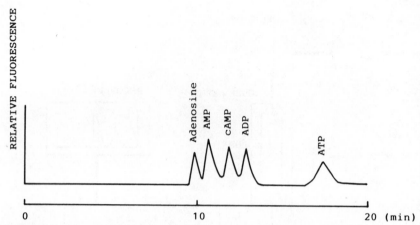

**FIGURE 6-5.** Chromatogram of a standard mixture of adenosine and adenine nucleotides detected by post-column derivatization with bromoacetaldehyde. Column: 4.6-mm ID × 35-mm length, packed with Hitachi gel 3012-N (anion exchange porous polystyrene polymer beads; 7 μm). Column temperature: 45°C. Mobile phase: 0.025 M citric acid-0.05M disodium hydrogen phosphate-0.4M sodium chloride (pH 5.0)/acetonitrile (4/1) containing 0.1 M bromoacetaldehyde. Flow rate: 0.1 ml/min. Post-column reaction coil: 0.1 mm ID × 30 m length. Heating bath temperature: 100°C. Fluorescence detection: excitation, 253.7nm; emission, 400nm. Peaks: AMP, adenosine monophosphate; cAMP, cyclic AMP; ADP, adenosine diphosphate; ATP, adenosine triphosphate. Each component present as 1 pmol.

ity. Using equations 19, 20, and 21, for $d_c$ equal to 1.5 mm, $F$ equal to 3.333 × $10^{-3}$ cm³/sec, which are typical values for the internal diameter of the reactor column and the total flow rate in semi-micro HPLC, assuming that ε equals 0.48, A equals γ equals 0.8, $D_M$ equals $10^{-5}$ cm²/sec, η equals $10^{-2}$ g/cm × sec², and $k_0$ equals 2 × $10^{-3}$, we obtain

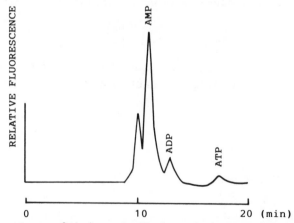

**FIGURE 6-6.** Chromatogram of adenine nucleotides in a rat brain. See Fig. 6-5 for conditions.

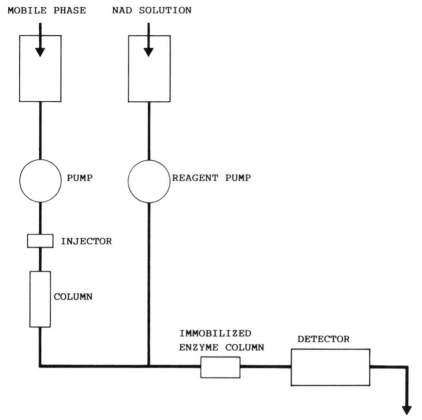

**FIGURE 6-7.** Flow diagram of a 3α- and 3β-hydroxysteroid analysis system using enzymatic post-column reaction.

$$t = 2.543L \tag{22}$$

$$\sigma_{tr}^2 = (2.631 \times 10^{-4} + 168.3d_p^{1.33})L \tag{23}$$

$$\Delta P = \frac{1.996L}{d_p^2} \tag{24}$$

If the same separation column is used, with the same allowance for $\sigma_{t(ob)}/\sigma_{tc}$ and $\Delta P$ as in the open-tubular reactors (shown in Table 6-2) for the peak with a capacity factor $k'$ equal to 1, the maximum reaction times $(t_{max})$ and the maximum reactor column lengths $(L_{max})$ can be calculated for different particle sizes, using equations 14, 22, 23, and 24. As shown in Fig. 6-2, the relationship between the reaction time, the reactor column length, and the peak broadening caused by the reactor can also be obtained under the same conditions as in the open-tubular reactors.

As mentioned so far, for semi-microcolumns, open-tubular reactors are suitable for reaction times smaller than ten and several seconds, while

Hydroxysteroids

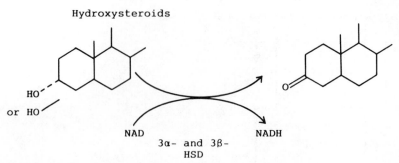

**FIGURE 6-8.** Enzymatic reaction of hydroxysteroids with NAD.

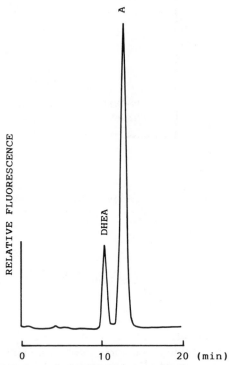

**FIGURE 6-9.** Chromatogram of a standard mixture of androsterone and dehydroepiandrosterone using enzymatic post-column reaction with NAD. Column: 1.5-mm ID × 250-mm length, packed with JASCO Fine SIL $C_1$-5. Column temperature: 25°C. Mobile phase: methanol/water (60/40). Flow rate: 0.1 ml/min. Immobilized enzyme column: 1.5-mm ID × 30-mm length, packed with μS-Enzymepak-HSD. Reagent: 0.3 mM NAD + 10 mM $K^+$ phosphate + 1mM EDTA + 0.05% 2-mercaptoethanol (pH 7.8). Flow rate: 0.1 ml/min. Temperature: 25°C. Fluorescence detection: excitation, 340 nm; emission, 470nm. Peaks: A, androsterone, DHEA, dehydroepiandrosterone. Amounts: 7.3 ng for A; 7.2 ng for DHEA.

packed-bed reactors are suitable for reaction times between ten and several seconds and one hundred and several tens of seconds.

## 6.3. Applications

Yoshioka and co-workers[4] developed an adenine nucleotide analyzer with post-column derivatization using microscale HPLC. Figure 6-3 shows a flow diagram of the analyzer. A post-column derivatization reagent, bromoacetaldehyde, is added to the eluent in advance. The column effluent, containing separated adenine nucleotides, and the reagent are introduced into the reaction coil, which is kept at 100°C, where a fluorescent reaction takes place. In the reaction, as shown in Fig. 6-4, strongly fluorescent products, 1-$N^6$-ethanoadenine ($\epsilon$-adenine) nucleotides, are yielded. The reaction products are then continuously monitored by a fluorometric detector with the

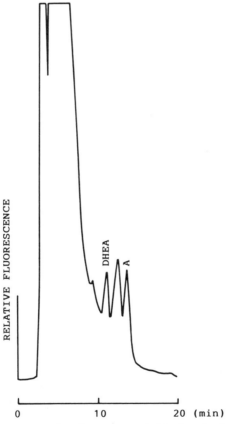

**FIGURE 6-10.** Chromatogram of androsterone and dehydroepiandrosterone in a normal human urine. For conditions see Fig. 6-9.

excitation wavelength at 253.7 nm and the emission wavelength at 400 nm. Figure 6-5 shows a chromatogram of a standard mixture of adenosine, AMP, cAMP, ADP, and ATP. Figure 6-6 shows a chromatogram of adenine nucleotides in a rat brain.

Takeuchi and co-workers[5] applied this method using an immobilized enzyme column as a post-column reactor[6] for microscale HPLC analysis of bile acids. This method can also be applied to the microscale HPLC analysis of 3α- and 3β-hydroxysteroids. Figure 6-7 shows the flow diagram of the analysis system. The column effluent containing separated hydroxysteroids is mixed with the NAD solution, which is delivered by a reagent pump. The effluent is introduced into an enzyme column where both 3α- and 3β-hydroxysteroid dehydrogenases are immobilized. In the enzyme column, hydroxysteroids are oxidized and NAD is reduced to NADH, which has a strong fluorescence, as shown in Fig. 6-8. Hydroxysteroids can be determined by monitoring NADH at 340 nm for excitation and at 470 nm for emission. Figure 6-9 shows the chromatogram of a standard mixture of androsterone (3α-hydroxysteroid) and dehydroepiandrosterone (3β-hydroxysteroid). Figure 6-10 shows a chromatogram of androsterone and dehydroepiandrosterone in a normal human urine.

### References

1. Deelder, R.S.; Kroll, M.G.F.; Van den Berg, J.H.M. *J. Chromatogr.* **1976, 125,** 307.
2. Deelder, R.S.; Kroll, M.G.F.; Beeren, A.J.B.; Van den Berg, J.H.M. *J. Chromatogr.* **1978, 149,** 669.
3. Deelder, R.S.; Kuijpers, A.T.J.M.; Van den Berg, J.H.M. *J. Chromatogr.* **1983, 255,** 545.
4. Yoshioka, M.; Tamura, Z.; Senda, M.; Miyazaki, T. *J. Chromatogr.* **1985, 344,** 345.
5. Takeuchi, T.; Saito, S.; Ishii, D. *J. Chromatogr.* **1983, 258,** 125.
6. Okuyama, S.; Kokubun, N.; Higashidate, S.; Uemura, D.; Hirata, Y. *Chem. Lett.* **1979, 12,** 1443.

# Applications of Microscale HPLC

## *D. Ishii, T. Takeuchi, and K. Hibi*

### *7.1. Introduction*

As compared with conventional HPLC, which uses columns with 4 to 8 mm diameters, micro-HPLC has the following advantages:

1. Low running cost due to a reduction in solvent consumption
2. High mass sensitivity due to low dispersion in the column
3. High resolution on a long column
4. Capability of new detection methods due to the low flow rate

Because of these features, HPLC has been utilized in many fields and numerous applications have been developed. The preceding chapters already illustrated some applications of micro-HPLC. Additional interesting applications of microscale HPLC in various fields are summarized in this chapter. This discussion will be divided into subchapters dealing with the application of micro-HPLC, semi-micro-HPLC, and high-speed HPLC.

### *7.2. Applications of Micro-HPLC*

### *7.2.1. Basic Investigations*

One of the features of micro-HPLC that uses PTFE tubing column is short column preparation time. This advantage is utilized in the investigation of the properties of packing materials and optimization of the analytical conditions. Hara *et al.* utilized micro-HPLC for the study of the optimiza-

*D. Ishii and T. Takeuchi* Department of Applied Chemistry, Faculty of Engineering, Nagoya University, Chikusa-ku, Nagoya 464, Japan.    *K. Hibi* Japan Spectroscopic Company, Ltd., Hachioji City, Tokyo 192, Japan.

tion of the solvent system in the separation of indole alkaloids by normal-phase HPLC using a silica gel column[1] and Kimura *et al.* utilized micro-HPLC for the selection of the best porous polystyrene gel for the separation of polymyxin antibiotics.[2] In an attempt to explain the hydrophobicity of drugs by their partition coefficients between *n*-octanol and water, Miyake *et al.* calculated the partition coefficients of drugs using micro-HPLC with a 0.5 mm ID PTFE microcolumn packed with silica gel coated with octanol.[3]

### 7.2.2. Analysis of Bile Acids in Serum

Analysis of bile acids in serum is required for diagnostic purposes because the abnormal presence of bile acids reflects a functional disorder of the liver. The increased mass sensitivity of micro-HPLC is significant in the analysis of biological samples.

HPLC analysis of bile acids with 3α-hydroxysteroid dehydrogenase (3α-HSD) post-column derivatization[4,5] seems promising with respect to resolution, sensitivity, and quantitation, as compared with GC, GC-MS, thin-layer chromatography (TLC) and HPLC using ultraviolet and refractive index detectors. In the method using a post-column enzyme reaction, the 3α-hydroxy group in each bile acid is oxidized to a keto group, while simultaneously β-NAD is reduced to NADH, which is subjected to fluorescence detection.

Figure 7-1 shows the block diagrams of micro-HPLC systems used for the analysis of bile acids. The upper system requires two pumps, one of which is supplying the reagent (NAD). The lower system employs a mobile phase containing NAD, and requires only a single pump, leading to increased sensitivity due to the elimination of pulsation.[6]

Sensitivity for bile acids is strongly affected by the pH of the mobile phase; a pH of around 10 gives the highest sensitivity.[7] However, a mobile phase with a pH around 9 was used, considering the stability of the separation column and NAD.

A

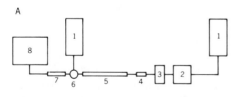

B

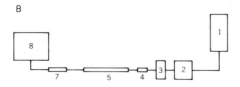

**FIGURE 7-1.** Block diagram of micro-HPLC systems for the analysis of bile acids. (A) Post-column mixing system. (B) Pre-mixing system. 1, Pump (Microfeeder); 2, gradient equipment; 3, microvalve injector; 4, guard column; 5, separation column; 6, T-piece; 7, immobilized enzyme column; 8, fluorometer.

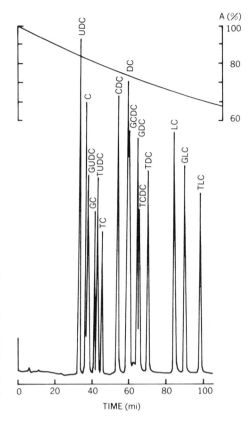

**FIGURE 7-2.** Gradient-elution separation of bile acids. Column: 0.26-mm ID × 200-mm length, packed with 5 μm silica-ODS particles. Mobile phase: (A) acetonitrile/60 mM phosphate solution (pH: 9.8)/60 mM phosphate solution (pH: 8.9) containing 18 mM NAD (20/70/10); (B) acetonitrile/60 mM phosphate solution (pH: 9.5)/60 mM phosphate solution (pH: 8.9) containing 18mM NAD (60/30/10). Flow rate: 2.1 μl/min. Sample volume: 0.011 μl. Peaks: C, cholic acid; GC, glycocholic acid; TC, taurocholic acid; DC, deoxycholic acid; GDC, glycodeoxycholic acid; TDC, taurodeoxycholic acid; CDC, chenodeoxycholic acid; GCDC, glycochenodeoxycholic acid; TCDC, taurochenodeoxycholic acid; LC, lithocholic acid; GLC, glycolithocholic acid; TLC, taurolithocholic acid; UDC, ursodeoxycholic acid; GUDC, glycoursodeoxycholic acid; TUDC, tauroursodeoxycholic acid.

Figure 7-2 shows the separation of a bile acids mixture on an ODS column. The mobile phase contains 1.8 mM NAD. The bile acids (approximately 20 ng each) are detected by a fluorescence detector and the sensitivity is increased by a factor of about 50, as compared to conventional HPLC. The detection limit (signal-to-noise ratio: 2) is 0.13–0.28 pmole.

Bile acids are present in serum at low concentrations. The micro precolumn concentration method is effective in the analysis of dilute samples. The details of the pre-column concentration method are described in Chapter 3. Bile acids in serum could be effectively concentrated onto the ODS precolumn by tenfold dilution of the serum with a phosphate buffer solution (pH: 7–8).[7] Figure 7-3 shows the linear relationships between the peak height and the concentration of bile acids.

Figures 7-4 and 7-5 illustrate the separation of bile acids in 0.1 ml serum of a healthy volunteer and of a patient with alcoholic cirrhosis.[7] The upper chromatograms were obtained with a post-column, in which fluorescent compounds other than NADH were detected. The difference in the amounts of bile acids between the two lower chromatograms is significant.

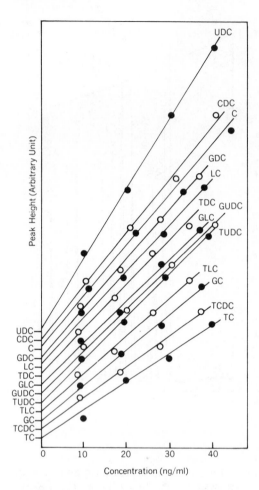

**FIGURE 7-3.** Relationship between peak height and the concentration of bile acids. Sample volume: 1 ml. Precolumn: 0.2-mm ID × 10-mm, packed with 15–30 μm silica-ODS particles. Abbreviations defined in Fig. 7-2 caption.

Figure 7-6 shows the separation of bile acids in calf serum.[8] Many peaks other than bile acids can be seen in the chromatogram.

### 7.2.3. Analysis of Amino Acids in Soy Sauce and in Sake

Liquid chromatographic analysis of amino acids has been investigated by many researchers. Derivatization with 5-dimethylaminonaphthalenesulphonyl (Dns or Dansyl) chloride is known as one of the very sensitive HPLC methods for amino-acid analysis.

Figure 7-7 demonstrates the gradient elution separation of a mixture of Dns-amino acids on a 10-cm long ODS column.[9] The detection limit (signal-to-noise ratio: 2) was approximately 0.1 pmole.

Figures 7-8 and 7-9 show gradient elution separations of amino acids in soy sauce and in sake, respectively.[9] Dansyl derivatization of amino acids in these samples was carried out at 38°C for 1–2 hr after adjusting the pH

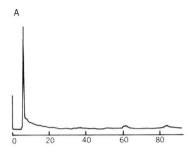

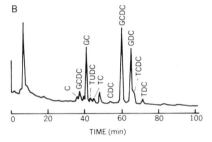

**FIGURE 7-4.** Separation of bile acids in serum of a healthy volunteer. Operating conditions as in Figure 7-2, except for the sample volume. Sample volume: 0.1 ml. (A) Operation without a post-column; (B) operation with a post-column. Peaks: as in Fig. 7-2.

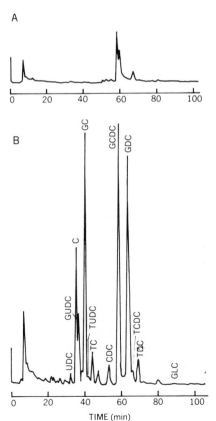

**FIGURE 7-5.** Separation of bile acids in the serum of a patient with alcoholic cirrhosis. Operating conditions as in Figure 7-2, except for the sample volume. Sample volume: 0.1 ml. (A) Operation without a post-column; (B) operation with a post-column. Peaks: as in Fig. 7-2.

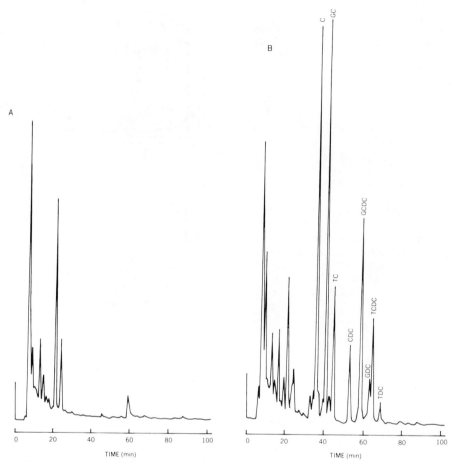

**FIGURE 7-6.** Separation of bile acids in calf serum. Operating conditions as in Figure 7-2, except for the sample. Sample: 0.1 ml of calf serum. (A) Without a post-column; (B) with a post-column. Peaks: as in Fig. 7-2.

of the sample solution to 9.7. The high concentration of the amino acids in these samples permitted injection of a small sample volume (0.02 $\mu$l) with a valve injector. The injected sample amounts corresponded to 1.3 nl of soy sauce and 11 nl of sake.

### 7.2.4. Oligomers

The byproducts of epoxy resin production are compounds having functional groups other than epoxide groups, either as end groups or as side chains. As these byproducts affect the property of epoxy resins, their characterization is of practical importance. Size-exclusion chromatography, reversed-phase liquid chromatography, any field-desorption mass spectromerty have been utilized to characterize epoxy resin oligomers. Separation of

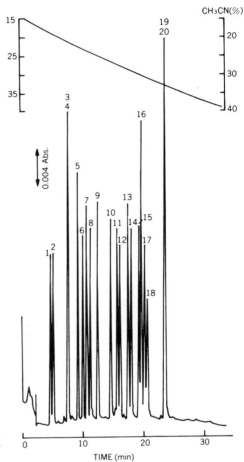

FIGURE 7-7. Gradient elution separation of dansyl-amino acids. Column: 100-mm × 0.34-mm ID, packed with ODS-Hypersil (3 μm). Mobile phase: acetonitrile/0.13M ammonium acetate, gradient profile as indicated. Flow rate: 4.2 μl/min. Wavelength of UV detection: 222 nm. Peaks: 1, Asp; 2, Glu; 3, Hyp; 4, Asn; 5, Ser; 6, Thr; 7, Gly; 8, Ala; 9, Pro; 10, Val; 11, Nval; 12, Met; 13, Ile; 14, Leu; 15, Nleu; 16, Trp; 17, Phe; 18, NH$_2$; 19, di-Cys; 20, Cys. Each peak corresponds to about 40 pmol of substance.

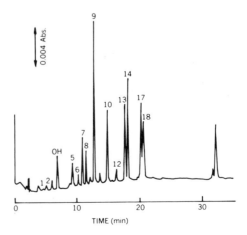

FIGURE 7-8. Separation of amino acids in soy sauce. Operating conditions and peak identification as in Fig. 7-7. OH, Dansylic acid.

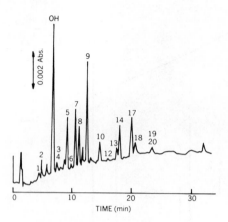

**FIGURE 7-9.** Separation of amino acids in sake. Operating conditions and peak identification as in Fig. 7-7. OH, Dansylic acid.

the epoxy resin oligomers by micro-HPLC was investigated in both the reversed-phase and size-exclusion modes.

Micro-HPLC facilitates the use of long columns providing large theoretical plate numbers. Figure 7-10 shows the separation of epoxy resin (Epikote 828) obtained on 50 cm and 2 m micro-SEC columns packed with 5 $\mu$m polystyrene gel having the exclusion limit of $2 \times 10^4$ molecular weight

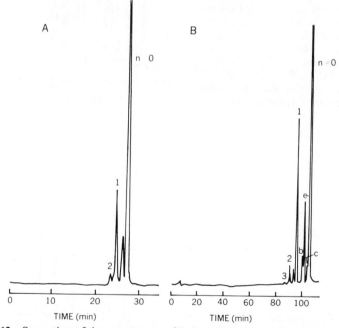

**FIGURE 7-10.** Separation of the components of Epikote 828. Columns: (A) 0.35-mm ID × 0.5-m packed with 5-$\mu$m polystyrene gel having the exclusion limit of $2 \times 10^4$ mol. wt.; (B) four 0.5-m columns connected in series. Mobile phase: tetrahydrofuran. Flow rate: 1.04 $\mu$l/min. Detection wavelength: 280 nm. Peaks: n = 0–3, oligomer of main product; b–e, byproducts.

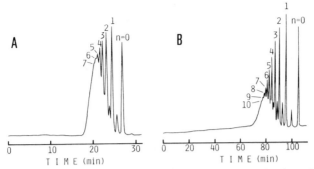

**FIGURE 7-11.** Separation of Epikote 1001. Operating conditions as in Figure 7-10. Peaks: n = 0–10, epoxy oligomers.

as polystyrene standard.[10] In Fig. 7-10B three byproduct peaks appear between the main peaks (n = 0 and n = 1). The flow rate is about 1 $\mu$l/min, consuming only 120 $\mu$l of the mobile phase. It is relatively easy to prepare and utilize such a long column in micro-SEC.

Figure 7-11 shows the separation of the components of epoxy resin (Epikote 1001) obtained under the same conditions as used for Fig. 7-10.[10] Small peaks corresponding to byproducts can be clearly observed between main peaks ranging from n = 0 to n = 10 on the chromatogram obtained on the 2 m long column, while the shorter column gives only limited separation.

The molecular weight calibration plot for epoxy resin oligomers, shown in Fig. 7-12, indicates a linear relationship between the logarithm of the molecular weight and the retention time, except for the n = 0 peak.[11]

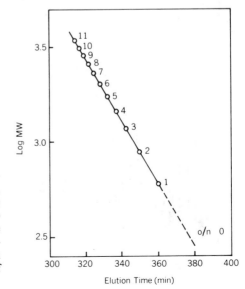

**FIGURE 7-12.** Molecular weight calibration plot. Column: 4-m × 0.33-mm ID, packed with polystyrene gel having the exclusion limit 6 × 10⁴ mol. wt. of polystyrene (TSK-GEL G3000H). Mobile phase: THF. Flow rate: 0.56 $\mu$l/min. Sample: Epikote 1001. The numbers 1 through 11 indicate the polymerization degree of each epoxy oligomer.

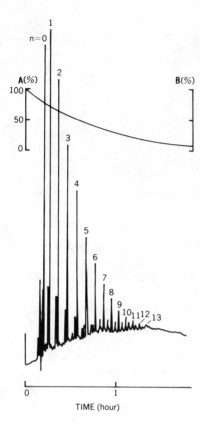

**FIGURE 7-13.** Gradient-elution separation of Epikote 1001. Column: 0.22-mm ID × 0.5-m, packed with 5-μm silica-ODS particles. Mobile phase: (A) acetonitrile/water (85/15); (B) acetonitrile/tetrahydrofuran (90/10), gradient profile as indicated. Flow rate: 1.4 μl/min. Sample amount: 0.16 μg. Detection wavelength: 225 nm.

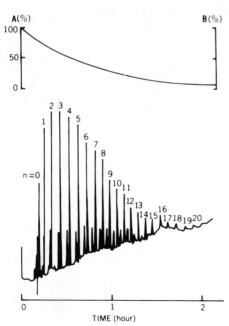

**FIGURE 7-14.** Gradient-elution separation of epoxy resin (Epikote 1004). Operating conditions as in Fig. 7-13. Gradient profile as indicated. Sample amount: 0.2 μ g.

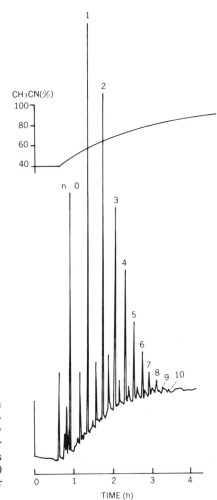

**FIGURE 7-15.** Separation of Epikote 1001 on an open-tubular column. Column: 22-m × 31-μm ID glass capillary which has chemically bonded ODS stationary phase on the inner surface. Mobile phase: acetonitrile/water, as indicated. Flow rate: 0.52 μl/min. Sample: 160 ng. Column temperature: 44°C. Detector wavelength: 225 nm.

Reversed-phase liquid chromatography generally results in a much better resolution of the oligomers than SEC. The exponential gradient profile is effective to separate oligomers. Thus, a gradient system with a mixing chamber providing an exponential gradient profile can be successfully applied to the separation of epoxy resin oligomers.[12]

Figures 7-13 and 7-14 demonstrate gradient-elution separation of Epikote 1001 and 1004 on a 0.22-mm ID × 50-cm long ODS column. Many peaks corresponding to the byproducts that may have chlorine or hydroxyl groups are also resolved. The resolution obtained on this reversed-phase column is superior to that obtained on the micro-SEC columns.

Figure 7-15 shows the separation of Epikote 1001 using a 22-m long × 31-μm ID open-tubular ODS column.[13] Resolution obtained on this column is somewhat poorer than that of the 0.5 m packed column.

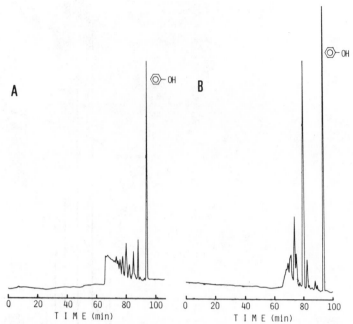

**FIGURE 7-16.** Separations of phenol-formaldehyde resin oligomers. (A) Resol; (B) Novolak, Column: four 0.35-mm ID × 0.5-m columns, packed with G1000H, in series. Mobile phase: THF. Flow rate: 1.04 μl/min.

Phenol-formaldehyde resins are produced by condensation or addition reactions between phenol and formaldehyde, and reaction conditions affect the structure of these resins. Linear polynuclear novolak and resol with numerous methylol groups are typical prepolymers of such resins. Since the structure and the sizes of species in phenol-formaldehyde resins are similar, resolution of the individual species is difficult.

Separation of phenol-formaldehyde resins in the size-exclusion mode using a 2-m long × 0.35-mm ID column packed with polystyrene gel having the exclusion limit of $1 \times 10^3$ molecular weight (TSK-GEL G1000H) is shown in Fig. 7-16.[10]

### 7.2.5. Determination of Organic Compounds in Water

Micro-HPLC using pre-column concentration method (see Chapter 3, Section 3.1.3) is effective for analysis of components present in water at low concentrations. Figure 7-17 demonstrates the gradient-elution separation of impurities present in distilled water in contact with powdered coal on a micro-packed fused-silica column.[12] Due to the high resolution of the column, a large number of peaks are resolved. Identification of these impurities is difficult, and may be solved by coupling high-resolution micro-HPLC and MS.

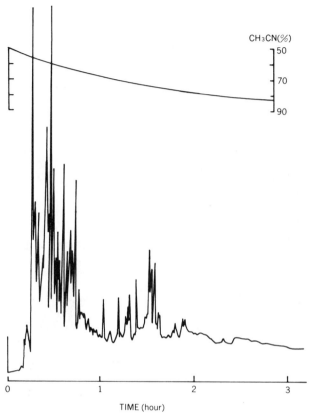

**FIGURE 7-17.** Gradient-elution separation of impurities present in distilled water in contact with powdered coal. Column: 0.5-m × 0.22-mm ID, packed with 5-μm silica-ODS particles. Mobile phase: acetonitrile/water, gradient profile as indicated. Flow rate: 1.04 μl/min. Sample: 2.5 ml of distilled water in contact with powdered coal. Pre-column: 10-mm × 0.2-mm ID packed with 15–30-μm silica ODS. Detector wavelength: 225 nm.

Another example of separation of the impurities present in deionized water is shown in Fig. 7-18.[14] The retention volume of the last peak coincided with that of dibutyl phthalate (DBP). An online high-pressure precolumn concentration system was applied to achieve the rapid analysis of DBP in water.[15] Figure 7-19 demonstrates the separation of DBP in tap water and commercially available purified water. The concentrations of DBP were found as 4.5 ppb and 5.4 ppb respectively.

### 7.2.6. Determination of Antioxidants in Gasoline

Gasoline contains various types of additives, which improve its performance. Antioxidants protect olefins and other unsaturated gasoline components from oxidation. The determination of such compounds is of prac-

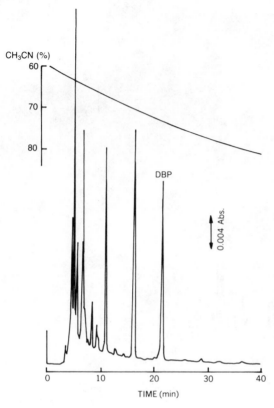

CH₃CN (%)

**FIGURE 7-18.** Gradient-elution separation of impurities present in deionized water. Column: 0.26-mm ID × 200-mm, packed with 5-μm silica-ODS particles. Mobile phase: Acetonitrile/water. Flow rate: 2.08 μl/min. Pre-column: 10-mm × 0.2-mm ID, containing 15–30-μm silica-ODS particles. Volume of mixing vessel: 109 μl. Sample: 3 ml deionized water. Detector wavelength: 235 nm.

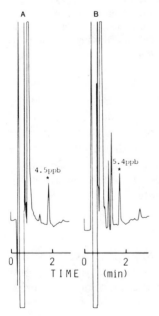

**FIGURE 7-19.** Separation of DBP in (A) tap and (B) purified water. Column: 100-mm × 0.34-mm ID, packed with 5 μm silica-ODS particles. Mobile phase: Acetonitrile/water (7/3). Pre-column: 27-mm × 0.34-mm ID, containing 15–30-μm silica-ODS particles. Flow rate: 20 μl/min. Sample volume: 0.7 ml. Detector wavelength: 235 nm. (A) Tap water; (B) purified water. Concentration of DnBP: (A) 4.5 ppb in tap water; (B) 5.4 ppb in purified water.

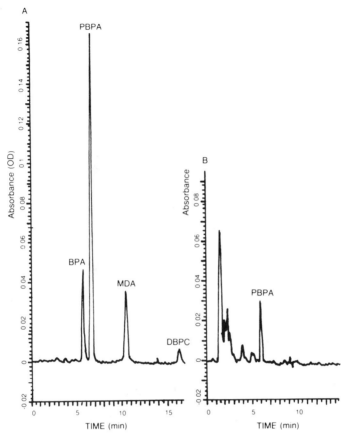

**FIGURE 7-20.** Separation of antioxidants and a metal deactivator. Column: 150-mm × 0.34-mm ID, containing 5-μm silica-ODS particles. Pre-column: 10-mm × 0.2-mm ID, containing 10-μm silica gel particles. Mobile phase: acetonitrile/water/n-hexylamine (65/35/1). Flow rate: 5.6 μl/min. Detector: Multi-channel UV detector. Wavelength: 290nm. Samples: (A) standard; (B) 1 μl gasoline. Peaks: BPA, N,N'-di-sec.-butyl-p-phenylenediamine; PBPA, N-phenyl-N'-sec.-butyl-p-phenylennediamine; MDA, N, N'-disalicylidene-1,2-propanediamine; DBPC, 2,6-di-tert.-butyl-p-cresol.

tical importance in estimating the stability of the gasoline and in controlling its quality.

The micro pre-column concentration method (see Chapter 3, Section 3.1.3) was applied to the analysis of antioxidants in gasoline.[16] Antioxidants such as p-phenylenediamines could be concentrated on alumina or silica gel pre-columns followed by separation in the reversed-phase mode. A multi-channel photodiode array detector is suitable for the measurement.

Figure 7-20 shows the separation of a mixture of various diamines and the components of gasoline.[17] The gasoline sample was diluted 300 times with n-hexane and 300 μl of the diluted sample was passed through a silica gel pre-column. Thus the chromatograms correspond to 1 μl gasoline.

Figure 7-21 illustrates a contour plot of the gasoline sample. Such a plot

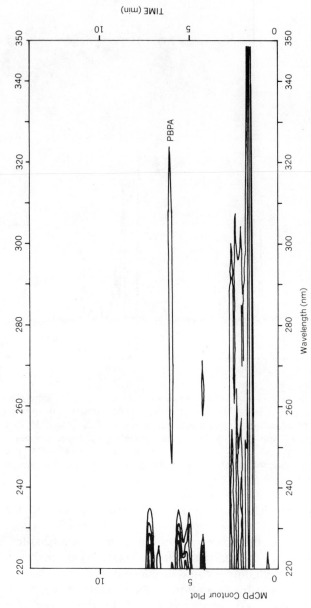

**FIGURE 7-21.** A contour plot of the gasoline sample. Operating conditions as in Fig. 7-20. PBPA = *N*-phenyl-*N'*-sec.-butyl-*p*-phenylenediamine.

is useful for selecting the optimal conditions for the determination by single-wavelength detection.

### 7.2.7. Determination of the Components of Medicines

The stepwise gradient separation of components in a cold medicine on a 4.9-m long × 52 μm ID open-tubular ODS column is demonstrated in Fig. 7-22. The medicine was dissolved in 10 ml of 5% aqueous acetonitrile solution and a 0.02 μl aliquot was injected.

Multichannel detection has many advantages over conventional single-wavelength spectrophotometers. Micro-HPLC using a photodiode array UV-visible detector was also applied to the analysis of components of a cold medicine.[18] Figure 7-23 shows the three-dimensional spectrochromatogram obtained. Each component in the figure could be identified by comparing its retention time and spectrum with those of a standard solute. Figure 7-24 illustrates the contour plot of the components.

### 7.2.8. Determination of Theophylline in Serum

High-performance liquid chromatography plays an important role in the determination of the concentration of various pharmaceuticals in blood. A typical measurement is the determination of theophylline, which is frequently used as a remedy for asthma. Because theophylline has a narrow range of efficient concentration in the blood and shows large differences in its metabolism between individuals, its concentration must be monitored in

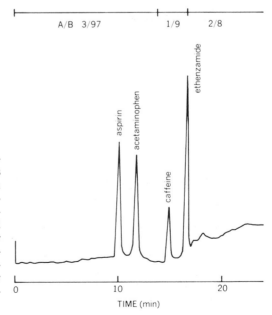

**FIGURE 7-22.** Stepwise gradient-elution separation of components present in a cold medicine on an open-tubular column. Column: 4.9-m × 52-μm ID glass capillary column, which has chemically bonded ODS stationary phase on the inner surface. Mobile phase: (A) acetonitrile; (B) 0.05% ammonium carbonate, composition as indicated. Flow rate: 1.1 μl/min. Detector wavelength: 225 nm.

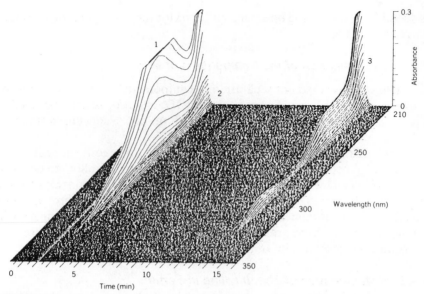

**FIGURE 7-23.** Three-dimensional spectrochromatogram of the components present in a cold medicine. Column: 0.34-mm ID × 100-mm, containing Hypersil-ODS (3 μm). Mobile phase: acetonitrile/methanol/60 mM phosphate solution (pH: 3.1; 14/6/80). Flow rate: 4.2 μl/min. Peaks: 1, acetaminophen; 2, caffeine; 3, ethenzamide.

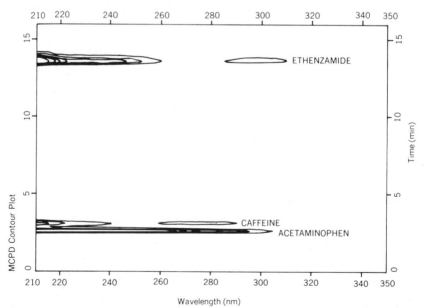

**FIGURE 7-24.** Contour plot of the components present in a cold medicine. Operating conditions as in Fig. 7-23. The plot is drawn from 0.02 absorbance to 0.3 absorbance at intervals of 0.03 absorbance.

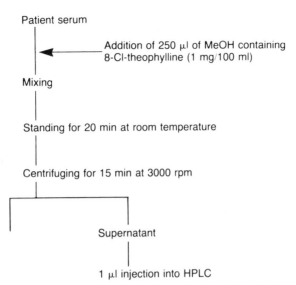

Patient serum

Addition of 250 µl of MeOH containing
8-Cl-theophylline (1 mg/100 ml)

Mixing

Standing for 20 min at room temperature

Centrifuging for 15 min at 3000 rpm

Supernatant

1 µl injection into HPLC

**FIGURE 7-25.** Pretreatment procedure for the determination of theophylline in serum.

the blood of each patient. Iwai and co-workers successfully measured the concentration of theophylline in blood serum by using micro-HPLC[19] with a short (15 cm) 0.5-mm ID tubing microcolumn consisting of PTFE tubing packed with 10 µm porous polystyrene gel, and 8-chlorotheophylline as the internal standard. They achieved a high accuracy; the relative standard deviation was below 1%. Figure 7-25 details the pretreatment procedure. Using this treatment, 99% of the protein present in the serum is removed and as much as 95% of the theophylline is successfully recovered. Figure 7-26 shows a typical chromatogram of the serum of a patient.

### 7.2.9. Combination of Micro-HPLC and Thin-layer Chromatography

Isolation by TLC is widely used as pretreatment for HPLC but, in many cases, the amount of sample fractioned and collected by TLC is very small. In such cases, micro-HPLC can be used very effectively because it allows the measurement of a small amount of sample with high sensitivity.

Morishita and co-workers succeeded in the analysis of obacunone and naringin contained in citrus fruits using the combination of micro-HPLC and TLC.[20] They first preseparated the crude limonoid extracted from defatted seeds of a summer orange by TLC, using dichloromethane as the mobile phase. They subsequently scratched off the spot considered as obacunone by comparison with a standard, extracted it using acetone, and removed the solvent. By using micro-HPLC to analyze the sample, they succeeded in identifying the obacunone component. Figure 7-27 shows a typical chromatogram.

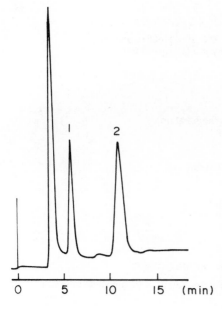

**FIGURE 7-26.** A typical chromatogram of the serum of a patient. Column: 0.5-mm ID × 15-cm, containing Fine GEL 110. Mobile phase: methanol/distilled water/acetic acid (70/30/1). Flow rate: 8 $\mu$l/min. Detector wavelength: 273 nm. Sample volume: 1 $\mu$l. Peaks: 1, theophylline; 2, 8-chlorotheophylline (internal standard).

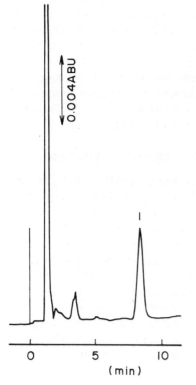

**FIGURE 7-27.** Analysis of obacunone after isolation by TLC. Column: 18-cm × 0.5-mm ID, containing Fine SIL $C_{18}$. Mobile phase: acetonitrile/0.1% aqueous phosphoric acid solution (50/50). Flow rate: 15 $\mu$l/min. Detector wavelength: 210 nm. Injection volume: 0.1 $\mu$l. Peak: 1, obacunone.

### 7.2.10. Applications of Fluorescence Detection in Micro-HPLC

Hibi and Kaneuchi developed a 0.6 $\mu l$ inner capacity microflow cell for a fluorescence detector, which can be used for micro-HPLC using a 0.5-mm ID PTFE tubing column. Ultra-high sensitivity analysis can be carried out by combining micro-HPLC with a high-sensitivity fluorescence detector.[21]

Figure 7-28 shows an example for the high-sensitivity analysis of poly-

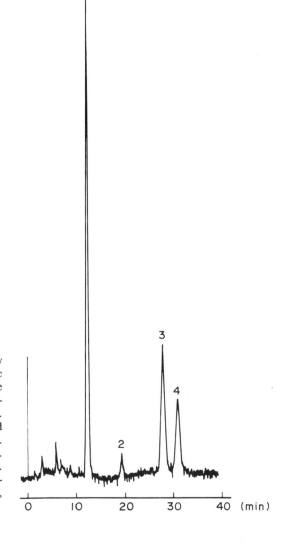

*FIGURE    7-28.* High-sensitivity analysis of polynuclear aromatic hydrocarbons with a fluorescence detector. Column: 16.7-cm × 0.5-mm ID, containing Fine SIL $C_{18}$. Mobile phase: acetonitrile/distilled water (80/20). Flow rate: 6 $\mu l$/min. Excitation wavelength: 254nm. Emission wavelength: 430 nm. Peaks: 1, anthracene (7pg); 2, chrysene (14pg); 3, perylene (9pg); 4, benz(a)pyrene (5pg).

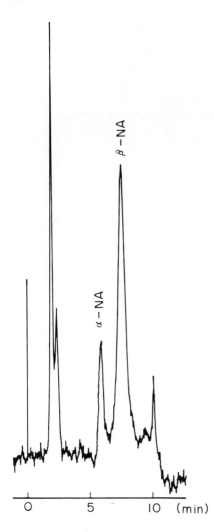

**FIGURE 7-29.** Separation and high-sensitivity detection of naphthylamine isomers. Column: 15-cm × 0.5-mm ID, packed with Fine SIL. Mobile phase: n-hexane/isopropanol(97/3). Flow rate: 15 µl/min. Excitation wavelength: 254 nm. Emission wavelength: 373 nm. Peaks: α-NA, α-naphthylamine; β-NA, β-naphtyhlamine.

nuclear aromatic hydrocarbons. As shown, extremely small amounts of sample components (5 to 14 pg) can be measured with a high signal-to-noise ratio.

Figure 7-29 shows the separation and high-sensitivity detection of isomers of naphthylamine, a carcinogen compound. Small amounts of α- and β-naphthylamine (1.7 pmole each) were separated on a microcolumn packed with 5 µm silica gel and detected by a fluorescence detector.

Figure 7-30 shows an example for the high sensitivity detection of vitamin $B_2$. Extremely small amounts (0.15 pmole) could be detected with a high signal-to-noise ratio.

Figure 7-31 shows an example of the high-sensitivity analysis of polyamine derivatives. Polyamines, biological amines that occur univer-

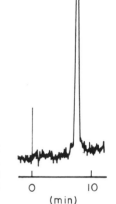

**FIGURE 7-30.** High-sensitivity detection of vitamin $B_2$. Column: 17.9-cm $\times$ 0.5-mm ID PTFE column, packed with Nucleosil $NH_2$. Mobile phase: acetonitrile/0.005 M phosphate buffer (75/25). Flow rate: 6 $\mu l$/min, Excitation wavelength: 365 nm. Emission wavelength: 530 nm. Sample: 0.15 pmol vitamin $B_2$.

sally in the biological matrix, have been attracting attention as an adjustment factor in the synthesis of protein or as an adjustment factor in the proliferation of cells. It has also been reported that the polyamine concentration increases in the urine of cancer patients. These observations make it very important to be able to analyze small amounts of polyamines in biological samples with high sensitivity. In this analytical example, the polyamines were converted into fluorescent substances by derivatizing the amino groups in the polyamine molecules using fluorescamine reagent. The fluorescent substances were subsequently separated and detected by micro-HPLC. Extremely small amounts of polyamines (0.2 pmole) could be detected.

### 7.2.11. Liquefied Solvents as the Mobile Phase

The viscosity of liquefied solvents is expected to be smaller than that of solvents commonly employed as the mobile phase in HPLC. Therefore the use of such solvents leads to higher column efficiency. Micro-HPLC permits the employment of low-boiling solvents, by keeping the pressure of the system sufficiently high to exceed the vapor pressure of the mobile phase. Diffusion coefficients of benzene in liquefied alkanes were calculated from Ouano's equation.[22] Table 7-1 lists these values.[23] The HETP is dependent on the diffusion coefficient and a solvent with a larger diffusion coefficient will result in a smaller theoretical plate height at the same flow rate. As dis-

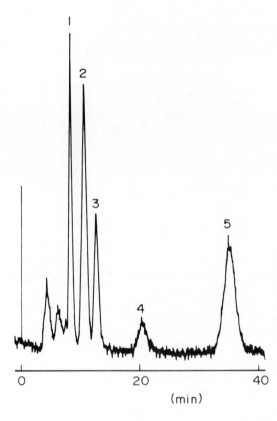

**FIGURE 7-31.** Separation and high-sensitivity detection of polyamine derivatives. Column: 0.5-mm ID × 16.8-cm, packed with LiChrosorb RP-8. Mobile phase: acetonitrile/0.1M boric acid (pH: 8.0; 29/71). Flow rate: 5 μl/min. Excitation wavelength: 365 nm. Emission wavelength: 485nm. Sample amount: 0.2 pmole each. Peaks: 1, putrescine; 2, cadaverine; 3, spermidine; 4, spermine; 5, 1,8-octanediamine (internal standard).

cussed in Chapter 3 (Section 3.2), mobile phases consisting of short-chain alkanes give better results in open-tubular LC.

Typical separations of PAHs using *n*-butane and propane as the mobile phase are shown in Fig. 7-32. Retention of solutes is dependent on the structure of the solvent as well as on its carbon number. A branched-chain sol-

**TABLE 7-1**

Diffusion Coefficients of Benzene in Alkanes

| Alkane | Diffusion coefficient $(cm^2/sec)$ |
|---|---|
| *n*-Hexane | $4.4 \times 10^{-5}$ |
| *n*-Pentane | $5.5 \times 10^{-5}$ |
| 2, 2-Dimethylpropane | $4.8 \times 10^{-5}$ |
| *n*-Butane | $6.9 \times 10^{-5}$ |
| 2-Methylpropane | $7.0 \times 10^{-5}$ |
| Propane | $1.0 \times 10^{-4}$ |

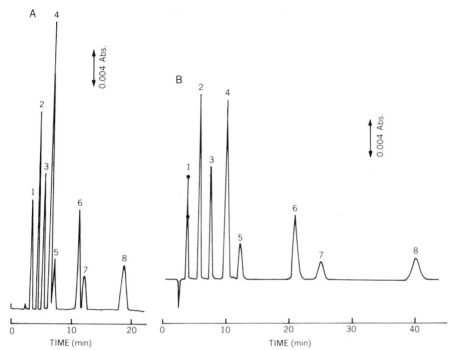

**FIGURE 7-32.** Typical separations of PAHs using short-chain alkanes as the mobile phase. Column: 0.5-mm ID × 125-mm, containing 5-μm silica gel (Develosil 100-5). Mobile phase: (A) *n*-butane; (B) propane. Flow rate: 10 μl/min. Detector wavelength: 254 nm. Peaks: 1, benzene; 2, naphthalene; 3, biphenyl; 4, anthracene; 5, pyrene; 6, chrysene; 7, benzo(a)pyrene; 8, 1,3,5-triphenylbenzene.

vent results in longer retention than a straight chain solvent with the same carbon number.

Figure 7-33 demonstrates the separation of aromatic amines and PAHs on a polystyrene column using liquefied dimethyl ether as the mobile phase.[24] The retention is primarily dependent on the number of aromatic rings in the solute molecules; solutes with more rings elute later.

For the analysis of polar solutes, binary solvent elution using two pumps was examined by Takeuchi and Ishii.[24] The composition was altered by changing the flow rate ratio of the two pumps. Experiments with low-boiling solvents could be carried out without any danger, since the mobile phase flow rate is very low in micro-HPLC.

### 7.2.12. Application to Supercritical Fluid Chromatography

Supercritical fluid chromatography (SFC) was first demonstrated experimentally by Klesper *et al.*[25] Supercritical fluids have liquidlike densities, but lower viscosities and higher diffusivities, resulting in a minimum plate height at higher linear velocities than in HPLC. *n*-Pentane and carbon diox-

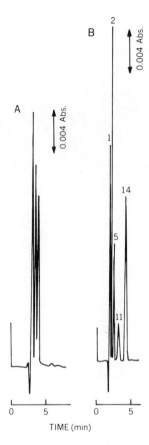

**FIGURE 7-33.** Separation of aromatic compounds using dimethyl ether as the mobile phase. Column: 103-mm × 0.26-mm ID, packed with polystyrene gel (TSK-GEL G1000H). Flow rate: (A) 1.4 μl/min; (B) 2.8 μl/min. Detector wavelength: (A) 235 nm; (B) 265 nm. Peaks: (A) aniline, α-naphthylamine, and N-phenyl-α-naphthylamine, eluting in this order. (B) 1, benzene; 2, naphthalene; 5, anthracene; 11, naphthacene; 14, benzo(a)pyrene.

ide have frequently been employed as the mobile phase in SFC. Carbon dioxide has many advantages compared with the solvents commonly employed in HPLC; it is nontoxic and nonflammable and has a good transparency at low UV wavelengths, a low critical temperature (approximately 31°C), and a low cost.

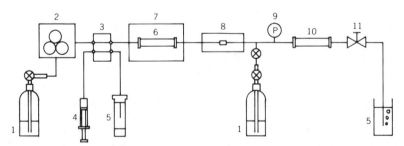

**FIGURE 7-34.** Diagram of the apparatus for supercritical fluid chromatography. 1, carbon dioxide cylinder; 2, pump; 3, microvalve injector; 4, sample introduction; 5, waste reservoir or drain; 6, separation column; 7, oven; 8, UV detector; 9, pressure gauge; 10, column for back pressure; 11, metering valve.

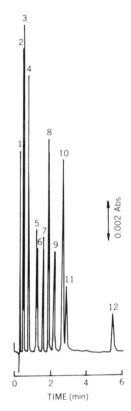

**FIGURE 7-35.** Separation of PAHs by supercritical fluid chromatography. Column: 150-mm × 0.34-mm ID, containing Silica-ODS (5 $\mu$m). Mobile phase: carbon dioxide. Inlet pressure: 150 kg/cm$^2$. Column temperature: 35°C. Detector wavelength: 245 nm. Sample: acetonitrile solution. Peaks: 1, benzene; 2, naphthalene; 3, biphenyl; 4, fluorene; 5, phenanthrene; 6, anthracene; 7, *p*-terphenyl; 8, 9-phenyl-anthracene; 9, fluoranthene; 10, 1,3,5-tri-phenylbenzene; 11, pyrene; 12, chrysene.

Figure 7-34 shows the diagram of the apparatus used for SFC.[26] A cylinder of carbon dioxide with a syphon supplies liquid carbon dioxide at ambient temperature. The pump head is cooled with solid carbon dioxide in order to improve pumping efficiency. The second carbon dioxide cylinder located downstream of the separation column is employed for priming carbon dioxide into the flow line between the separation column and the drain, but it could be eliminated by using tube fittings and low-dead-volume connecting tubing. The whole system, including the injector, columns, and detector flow cell is constructed to withstand high pressures.

Figure 7-35 demonstrates the separation of PAHs on a 150-mm long × 0.34 mm ID ODS column.[26] The inlet pressure was 150 kg/cm$^2$ and the column temperature was 35°C. The solute peak shapes are symmetric. Figure 7-36 shows the separation of styrene oligomers on the same column.[26] Although the detection wavelength was very low (205 nm), a smooth baseline is observed, due to the good transparency of carbon dioxide. Diastereoisomers are also separated in the chromatogram.

When pure carbon dioxide is used as the mobile phase, the separable solutes are restricted due to the polarity of CO$_2$. Therefore, a way should be developed for micro-SFC to incorporate a modifier into carbon dioxide.

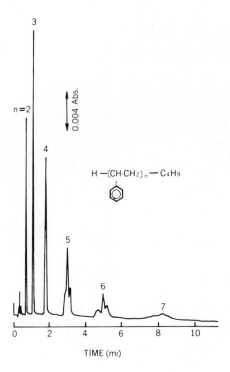

**FIGURE 7-36.** Separation of styrene oligomers by SFC. Column: as in Fig. 7-35. Mobile phase: carbon dioxide. Inlet pressure: 150 kg/cm². Column temperature: 34.5°C. Detector wavelength: 205 nm. Sample: Polystyrene A-500. The number above each peak gives the value of $n$ in the formula.

### 7.2.13. Other Applications

In addition to the applications discussed above, there are a large number of other applications mentioned in the literature. An example is the separation and direct detection of saccharides by a UV detector. Figure 7-37 shows an example of this method. The saccharides were detected at 190 nm and micro-HPLC was used without any pre-column derivatization. In the case of a flow cell having a light path length of 10 mm, which is used in conventional HPLC, a short wavelength such as 190 nm cannot generally be used because of the absorption of the acetonitrile present in the mobile phase. In micro-HPLC, however, the direct detection of saccharides is possible even at 190 nm because the absorption of the acetonitrile in the eluent is reduced to approximately $\frac{1}{30}$th of that observed in a conventional flow cell. This is due to the 0.3-mm light pathlength of the microflow cell (capacity 0.3 $\mu$l).

There has been a unique report on the application of micro-HPLC to the simultaneous quantitative analysis of trace elements such as lead, zinc, and copper in water. Yamazaki et al.[27] extracted the metal ions from water into a chloroform layer as the diethyldithiocarbamate chelates and then directly injected the extract into a micro-HPLC. Figure 7-38 shows a typical chromatogram. They applied this method to the analysis of metal ions in river water and reported that the values thus obtained were in good agreement with the results obtained by atomic absorption spectroscopy.

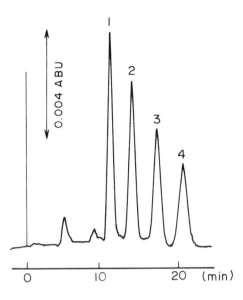

**FIGURE 7-37.** Separation of saccharides and their direct UV detection. Column: 15-cm × 0.5-mm ID, PTFE tube packed with Nucleosil NH₂. Mobile phase: acetonitrile/distilled water (70/30). Flow rate: 5 μl/min. Detection wavelength: 190 nm. Injection volume: 1.9 μl. Sample: 0.5% solution in the mobile phase. Peaks: 1, xylose; 2, glucose; 3, fructose; 4, lactose. Each peak corresponds to 9.5 μg.

## 7.3. Application of Semi-Micro-HPLC

### 7.3.1. Introduction

Semi-micro-HPLC, which uses a column with one-tenth of the volume of a conventional column, seems less effective in terms of cost and performance than real micro-HPLC, which uses a column with a column volume of less than one-hundredth of the volume of a conventional column. In

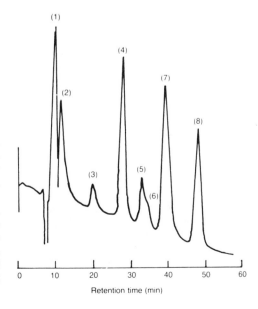

**FIGURE 7-38.** Chromatogram of the extract containing Cd(II)-, Mn(II)-, Pb(II)-, Co(II)-, Fe(III)-, Zn(II)-, and Cu(II)-diethyldithiocarbamate chelates. Column: 15-cm × 0.5-mm ID, containing 5-μm Silica-ODS particles. Mobile phase: methanol/ethyl acetate/water/ 0.05M Na-diethyldithiocarbamate solution (66/5/24/5). Flow rate: 4 μl/min. Detection wavelength: 265 nm. Injection volume: 0.4 μl. Peaks and the corresponding metal concentration in the aqueous solution: 1, Cd (5 μg/ml); 2, chloroform; 3, Mn (0.5 μg/ml); 4, Pb (6 μg/ml); 5, Co (5 μg/ml); 6, Fe (2 μg/ml); 7, Zn (4 μg/ml); 8, Cu (2 μg/ml).

practical HPLC measurement of samples, however, it has the following advantages:

1. Semi-micro-HPLC is almost as easy to operate as conventional HPLC. On the other hand, measurement by real micro-HPLC requires special skills, techniques, and great care.
2. Both the gradient-elution method and the post-column derivatization method are applicable.
3. Column and HPLC systems are commercially available from several manufacturers.

These three advantages are indispensable for the routine use of HPLC as an analytical tool. They are also extremely important in upgrading the analytical HPLC system from the conventional type to the semi-micro type. A study of recent trends shows that the application of semi-micro-HPLC is rapidly expanding in various fields, and this is considered to be due largely to these three factors. Among the applications of semi-micro-HPLC reported thus far, some interesting examples will be shown in this section.

Another important point to consider when upgrading an analytical HPLC system to a semi-micro type is the performance of the semi-micro-column. Figure 7-39 compares the performance of a 1.5-mm ID semi-micro-

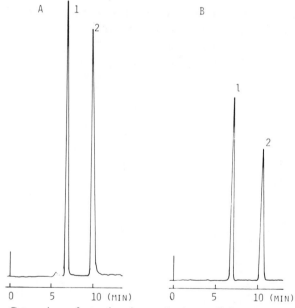

**FIGURE 7-39.** Comparison of retention time and column efficiency. (A) Semi-micro HPLC. Pump: FAMILIC-300S. Detector: UVIDEC-100-III. Column: μS-Finepak SIL C$_{18}$. Column size: 25-cm × 1.5-mm ID. Mobile phase: acetonitrile/water (75/25). Flow rate: 100 μl/min. Peaks: 1, naphthalene; 2, anthracene. (B) Conventional HPLC. Pump: TRI ROTAR V. Column: Finepak SIL C$_{18}$. Column size: 25-cm × 4.6-mm ID. Flow rate: 1.0 ml/min. Other conditions as in A.

column and a 4.6-mm ID conventional column. To apply the analytical conditions of conventional HPLC directly to semi-micro-HPLC, when the same packing material is used, it is essential that the semi-microcolumn should provide the same efficiency as the conventional column and the capacity factors should be equivalent on both columns. If this is not guaranteed, all the analytical conditions will have to be reexamined when the scale of HPLC is changed from conventional into semi-micro. The time required for such reexamination would present a serious problem in laboratories where HPLC is used for routine analysis.

In the example shown in Fig. 7-39, reversed-phase packing material was used. Concerning other typical packing materials, such as silica gel, aminopropyl silica gel, and porous polystyrene gel, our investigations have verified that a 1.5-mm ID semi-microcolumn provides a column efficiency and capacity factor equivalent to those obtained on a conventional column.

### 7.3.2. Analysis of Pharmaceutical and Biochemical Samples

Figure 7-40 shows an example for the analysis of anticonvulsants by semi-micro-HPLC. Phenobarbital, phenytoin, and carbamazepine were sep-

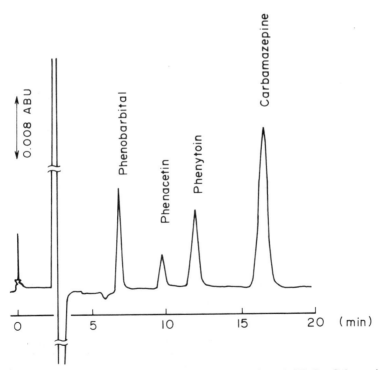

**FIGURE 7-40.** Separation of anticonvulsants. Column: $\mu$S-Finepak SIL $C_{18}$. Column size: 25-cm $\times$ 1.5-mm ID. Mobile phase: acetonitrile/0.005 M potassium dihydrogenphosphate in water (35/65). Flow rate: 150 $\mu$l/min. Detection wavelength: 220 nm.

arated using a 1.5 mm ID semi-microcolumn packed with silica with bonded $C_{18}$ groups. These drugs have a narrow range of efficacious concentrations in blood and large differences in metabolism between individuals. Moreover, excessive doses of these drugs produce various harmful side effects. The measurement of their concentrations in blood is therefore extremely important in establishing optimum dosage.

White and Laufer[28] analyzed cephalosporin antibiotics, using a column consisting of a 1.0-mm ID glass-lined stainless-steel tube containing bonded $C_{18}$ packing, and succeeded in separating five types of cephalosporin antibiotics. Figure 7-41 shows one of their chromatograms. They also analyzed cephalosporin in fermentation broths after a simple pretreatment consisting only of dilution, utilizing the high sensitivity of semi-microcolumn LC. White and Laufer also analyzed the impurities present in cephalosporin antibiotics using gradient elution.

Baram *et al.* constructed a semi-micro-HPLC system equipped with a UV detector permitting a rapid wavelength change according to a stepwise cyclic program.[29] They separated 14 types of amino acids using a 62 mm long × 2mm ID column packed with silica gel saturated with a copper-

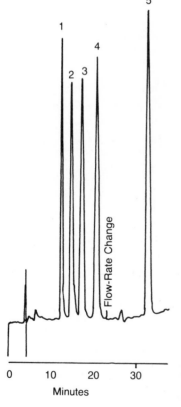

**FIGURE 7-41.** Separation of a mixture of cephalosporin antibiotics on a microbore column. Column: 1.0-mm ID × 25-cm, containing μ Bondapak $C_{18}$ (10 μm particles). Mobile phase: 0.01 M sodium dihydrogen phosphate/methanol (75/25). Flow rate: 50 μl/min (initially), 150 μl/min (after 23 min). Sample volume: 5 μl. Peaks: 1, cephalexine (0.05 μg); 2, cefoxitin (0.05 μg); 3, cephradine (0.07 μg); 4, cephaloglycin (0.10 μg); 5, cephalothin (0.23 μg).

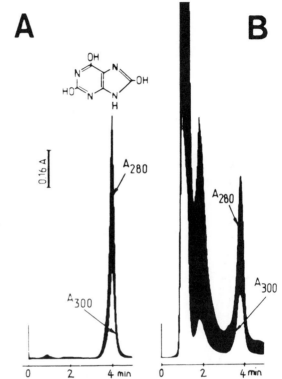

**FIGURE 7-42.** Direct analysis of uric acid in human serum by reversed-phase chromatography. Column: 62-mm × 2-mm ID, containing Nucleosil $C_{18}$ (5μm particles). Mobile phase: methanol/0.01M acetic acid in water (4/96). Flow rate: 100 μl/min. Detection wavelength: 280 and 300 nm. (A) Standard, 5 μg of uric acid in water; (B) 5 μl of normal human blood serum.

ammonium complex. They also successfully analyzed uric acid in serum; their chromatogram is shown in Fig. 7-42.

### 7.3.3. Applications of Gradient Elution

The gradient-elution method has been widely used in HPLC and it is now a common analytical technique. Two methods have been developed: the

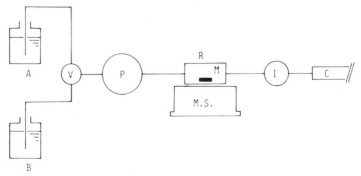

**FIGURE 7-43.** Block diagram of a gradient-elution system designed for semi-micro-HPLC. A, Initial solvent; B, final solvent; V, three-way switching valve; P, pump; M, stirrer; R, reservoir; M.S., magnetic stirring device; I, injector; C, column.

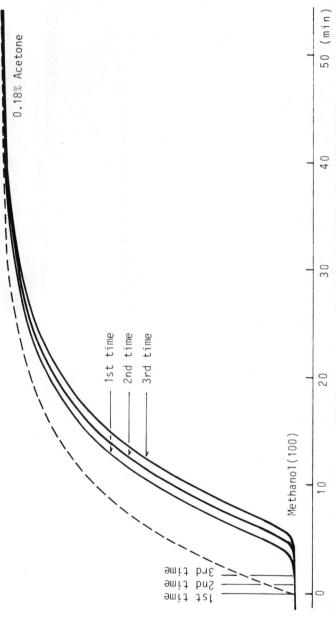

**FIGURE 7-44.** Comparison of the measured and calculated gradient profile. Solvent A: methanol. Solvent B: 0.18% acetone in methanol. Reservoir: 7.2-mm ID × 50-mm stainless-steel tube. Flow rate: 200 μl/min. Detection wavelength: 260 nm. Dotted line, calculated gradient profile; solid line, measured gradient profile.

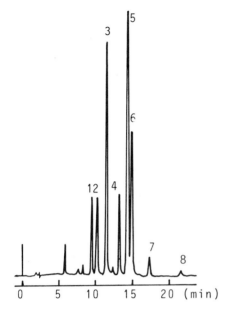

**FIGURE 7-45.** Chromatogram of PAHs using gradient elution. Column: $\mu$S-Finepak SIL $C_{18}$. Column size: 1.5-mm ID $\times$ 250-mm. Initial solvent: acetonitrile/water (50/50). Final solvent: acetonitrile (100). Flow rate: 150 $\mu$l/min. Reservoir: 7.2-mm ID $\times$ 50-mm stainless-steel tube. Peaks: 1, naphthalene; 2, biphenyl; 3, fluorene; 4, anthracene; 5, pyrene; 6, triphenylene; 7, chrysene; 8, benzo(a)pyrene. For reproducibility data see Table 7-2.

single-pump, low-pressure mixing gradient-elution method, and the two-pump, high-pressure mixing gradient-elution method. The single-pump gradient elution method is now used universally, because of its two advantages: it permits the addition of a ternary solvent and it has a low cost.

In micro-HPLC, gradient elution has been used to only a small extent, because it is very difficult to construct such a small-scale gradient device. Takahashi and co-workers have developed a simple device that permits single-pump gradient elution even in the flow-rate region of several tens of $\mu$l/min, representing the field of semi-micro-HPLC.[30]

**TABLE 7-2**

Reproducibility of Retention Time in Gradient Elution with a Semi-Microcolumn

| Peak number[a] | Average retention time (min)[b] | Standard deviation (min) | Relative standard deviation (%) |
|---|---|---|---|
| 1 | 9.36 | 0.084 | 0.90 |
| 2 | 10.10 | 0.078 | 0.77 |
| 3 | 11.40 | 0.088 | 0.77 |
| 4 | 13.14 | 0.096 | 0.73 |
| 5 | 14.24 | 0.108 | 0.76 |
| 6 | 14.80 | 0.106 | 0.72 |
| 7 | 17.30 | 0.106 | 0.61 |
| 8 | 21.62 | 0.154 | 0.71 |

[a]See figure 7-45.
[b]n = 6.

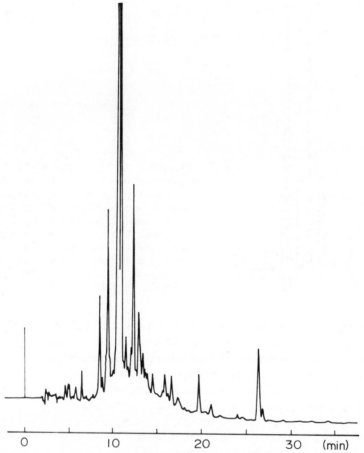

**FIGURE 7-46.** Analysis of "Kakkonto", a traditional Japanese herbal cold medicine. Column: μS-Finepak SIL $C_{18}$. Column size: 250-mm × 1.5-mm ID. Initial solvent: 0.1% phosphoric acid in water. Final solvent: acetonitrile/0.2% phosphoric acid in water (50/50). Volume of reservoir: 1.77 ml. Flow rate: 150 μl/min. Sample: extract of the cold medicine with hot water (60°C). Injection volume: 1.0 μl. Detection wavelength: 250 nm.

Figure 7-43 shows the block diagram of this gradient device. It consists of a three-way switching valve and a variable-volume high-pressure mixing chamber with a magnetic stirrer. After the whole system, including the mixing chamber and the column, have been purged with the initial solvent, the final solvent is delivered through the mixing chamber into the column by switching the three-way valve. Thus, the composition of the mixed solvent, flowing from the mixing chamber to the column varies exponentially until the initial solvent in the mixing chamber is replaced by the final solvent. The gradient curve is expressed in the following equation:

$$C_t = C_f - (C_f - C_i) \exp(-Ft/V_m) \qquad (1)$$

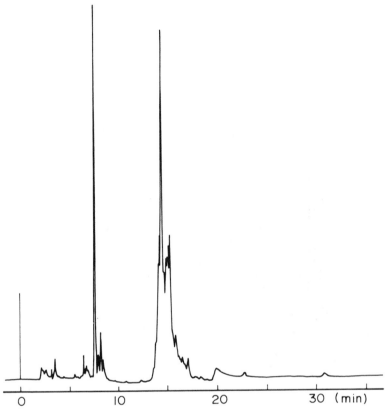

**FIGURE 7-47.** Separation of saponin components. Column: as in Fig. 7-46. Initial solvent: acetonitrile/0.1% phosphoric acid in water (10/90). Final solvent: acetonitrile/0.1% phosphoric acid in water (70/30). Flow rate: 150 µl/min, Detection wavelength: 220 nm. Sample: saponin compound (10 mg/ml) in water. Injection volume: 1.0 µl.

where $C_t$ is the concentration of mixed solvent at time t, $C_f$ is the concentration of final solvent, $C_i$ is the concentration of initial solvent, F is the mobile-phase flow rate, $V_m$ is the volume of mixing chamber, and t is the time. The gradient condition can be optimized by simply selecting the concentration of the initial and final solvents and the volume of the mixing chamber.

Figure 7-44 compares the actually measured gradient curve with the curve calculated using the above equation. The ordinate represents UV absorbance at 260 nm (0.32 AUFS). The gradient curve was measured by using methanol as the initial solvent and 0.18% acetone in methanol as the final solvent and direct connection of mixing chamber to UV flow cell. As shown in the figure, the delay time of gradient start is 3 min at a flow rate of 200 µl/min and the shape of the measured curve agrees with the calculated curve.

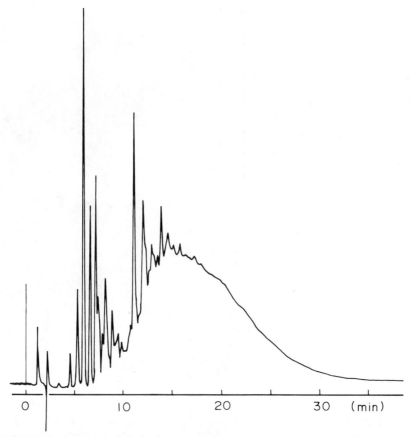

**FIGURE 7-48.** Analysis of a phenolic resin. Column: as in Fig. 7-46. Initial solvent: THF/distilled water (40/60). Final solvent: THF/distilled water (80/20). Detection wavelength: 280 nm. Sample: phenolic resin (24 mg/ml) in THF. Injection volume: 1.0 $\mu$l.

Figure 7-45 shows a typical chromatogram obtained in this system and Table 7-2 gives typical retention time reproducibility data. As seen the relative standard deviation of the retention time of eight aromatic hydrocarbons is better than 1%.

Figures 7-46 through 7-48 show applications of the gradient-elution system.[31] Figure 7-46 illustrates the analysis of "Kakkonto," a traditional Japanese herb cold medicine, using a 1.5-mm ID semi-microcolumn. The herb generally contains many components and gradient elution is indispensable for the HPLC analysis of all components. Figure 7-47 represents the separation of the components of saponin, which is used as an expectorant. Finally, Fig. 7-48 shows the analysis of a phenolic resin. The analysis of a phenolic resin is generally carried out by SEC; however, it is preferable to use reversed-phase HPLC to obtain detailed information on the low-molecular-weight components. On the other hand, in reversed-phase HPLC, the high-molecular-weight components present in the phenolic resin are

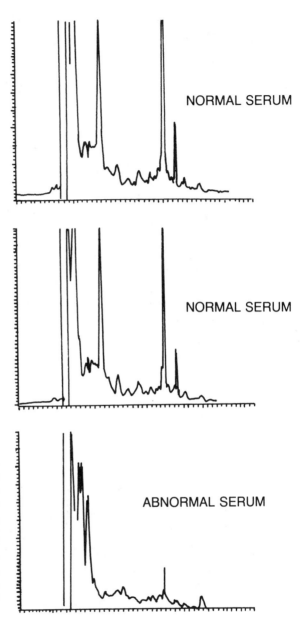

NORMAL SERUM

NORMAL SERUM

ABNORMAL SERUM

**FIGURE 7-49.** Chromatogram of blood serum samples separated on a microbore column. Column: 1-mm ID × 1-m containing a bonded $C_{18}$ packing (10-$\mu$m particles). Solvent: (A) methanol/water (75/25); (B) 100% methanol, exponential gradient (Curve No. 9, Waters Solvent Programmer). Flow rate: 40 $\mu$l/min. Detector wavelength: 254 nm.

strongly retarded by the column, and gradient elution is necessary to elute these components. In this analytical example, the low-molecular-weight components of the phenolic resin were separated with high resolution while the high-molecular-weight components were successfully eluted within a reasonable time.

The use of the high-pressure mixing gradient-elution method, utilizing

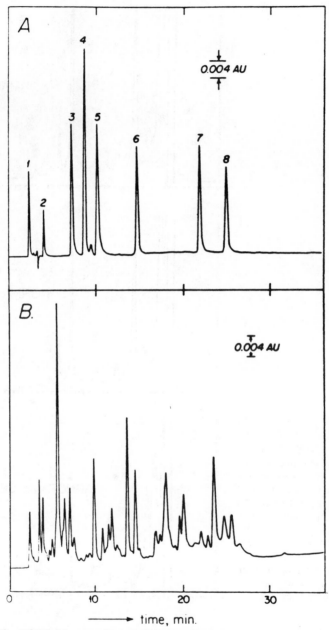

**FIGURE 7-50.** (A) Microbore gradient-elution separation of a peptide sample. Column: 1-mm ID × 35-cm, containing Zorbax-BP-ODS (7.5-$\mu$m particles). Solvent: acetonitrile/water containing 0.1% phosphoric acid with 10 mM $KH_2PO_4$ (5/95). Gradient: 10–100% B in 36 min. Flow rate: 80 $\mu$l/min. Detection wavelength: 214 nm. Sample volume: 5 $\mu$l. Peaks: 1, GlySer; 2, AlaVal; 3, PheGly; 4, TyrTyr; 5, AlaPhe; 6, ValAlaAlaPhe; 7, GlyPhePhe; 8, TrpTrp. (B) Microbore gradient-elution separation of insulin hydrolysate. Solvent: acetonitrile/water containing 0.1% phosphoric acid with 10 mM $KH_2PO_4$ (50/50). Sample volume: 50 $\mu$l. All other conditions as in A.

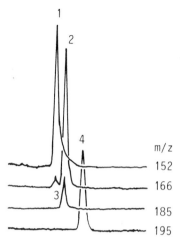

**FIGURE 7-51.** Mass fragmentogram of components present in commercial cold medicine. Column: $\mu$S-Finepak SIL C$_{18}$. Column size: 1.5-mm ID $\times$ 25-cm, packed with 10-$\mu$m silica-ODS particles. Mobile phase: methanol/water (90/10). Flow rate: 70 $\mu$l/min, Detector: MS-100, multi-channel selected ion monitoring. Peaks: 1, acetaminophen [M+H]$^+$; 2, ethenzamide [M+H]$^+$; 3, allyl isopropyl acetyl urea [M+H]$^+$; 4, caffeine, [M+H]$^+$.

two pumps has also been reported by several researchers.[28,32–35] Scott and Kucera studied the basic performance of such a system and reported on some of its applications.[34] Figure 7-49 represents an example. Here, blood serum samples were analyzed on a long microbore (1 mm ID) stainless-steel column. Schwartz *et al.* also studied the high-pressure mixing gradient-elution method, in which they attempted to apply the method for the analysis of peptides and proteins.[35] Figure 7-50 shows one of their applications. In this case, various peptides and insulin hydrolysate were analyzed on a 35 cm long microbore (1mm ID) column.

### 7.3.4. Combination of Semi-Micro-HPLC with Mass Spectrometry

We have studied the possibility of combining semi-micro-HPLC with MS in detail. Usually, flow rate of approximately 100 $\mu$l/min is used in semi-micro-HPLC with a 1.5 mm ID column. In the combination of semi-micro-HPLC and mass spectrometry, 70% and 90% of the effluent is removed by a splitter and 10% to 30% is introduced into the mass spectrometer through an interface. The authors utilized the vacuum nebulizing interface developed by T. Tsuge and described earlier in Chapter 5, Section 5.2.

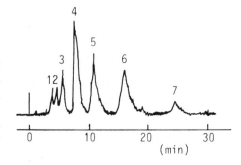

**FIGURE 7-52.** Analysis of fatty acids in the extract of saponified palm oil. Column: as in Fig. 7-51. Mobile phase: methanol/water (90/10). Flow rate: 80 $\mu$l/min. Detection: MS-100, total ion monitoring. Peaks: 1, caproic acid; 2, caprylic acid; 3, capric acid; 4, lauric acid; 5, myristic acid; 6, palmitic acid; 7, stearic acid.

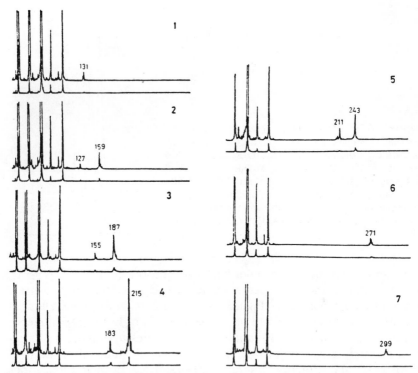

**FIGURE 7-53.** Mass spectra obtained at the individual peak tops in Fig. 7-52. Spectra identifications: 1, caproic acid; 2, caprylic acid; 3, capric acid; 4, lauric acid; 5, myristic acid; 6, palmitic acid; 7, stearic acid.

Figure 7-51 shows the mass fragmentogram obtained in the analysis of a commercially available cold medicine capsule. It indicates that even the components of an unresolved peak can be selectively detected by using multi-channel selected ion monitoring.

Figure 7-52 shows the analysis of fatty acids present in the extract of a saponified palm oil. The palm oil was saponified for one hour of 50°C in methanol containing 1N potassium hydroxide. The fatty acids thus saponified were extracted with chloroform and, after vaporization to dryness, the residue was dissolved in methanol and the methanol solution was analyzed by semi-micro-HPLC-MS. The individual peaks in the total ion chromatogram were identified from the mass spectra obtained at the individual peak tops. Figure 7-53 shows the corresponding mass spectra.

### 7.3.5. Other Interesting Applications

In addition to the applications described previously, the application of semi-micro-HPLC has also been reported in a number of other fields. For example, Kucera *et al.* reported on the separation of various samples, which are difficult to separate by conventional HPLC. Figure 7-54 shows the full separation of benzene and hexadeuterobenzene on a 4.5-m long × 1-mm

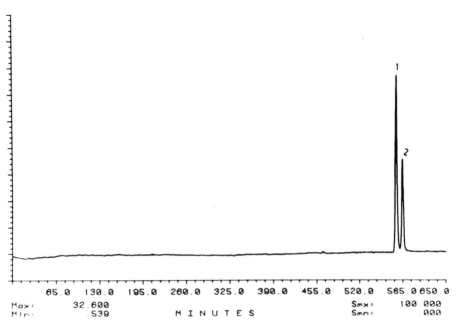

**FIGURE 7-54.** Chromatogram of a mixture of benzene and hexadeuterobenzene. Column: 1-mm ID × 4.5-m, packed with Zorbax ODS (8 μm). Mobile phase: methanol/water (85/15). Flow rate: 10 μl/min. Detection wavelength: 254 nm. Injection volume: 0.5 μl. Peaks: 1, hexadeuterobenzene; 2, benzene.

ID column packed with bonded $C_{18}$. It should be noted that the time required for the analysis was about 600 min.[36]

Jin and Rappaport applied electrochemical detection to microbore LC and succeeded in analyzing the nitro-substituted polynuclear aromatic hydrocarbons in diesel soot.[37]

Finally, an example of separating optical isomers by the use of a semi-microcolumn is shown. Figure 7-55 shows the chromatogram obtained in the analysis of D- and L-binaphthol on a column containing a chiral phase. This packing material, developed by Okamoto *et al.,* consists of a silica gel coated with the optically active polytriphenyl ethyl methacrylate.[38] Generally, the degree of separation of optical isomers is determined only by the properties of the packing material used and cannot be improved by changing the composition of the mobile phase. Therefore, the possibility of using a long semi-microcolumn, which has a higher efficiency, is very important in the field of optical isomer separation.

## 7.4. Application of High-Speed HPLC

### 7.4.1. Introduction

High-speed HPLC (fast HPLC), a technique that can drastically reduce the time required for analysis by conventional HPLC, is very important in laboratories where HPLC is used for the routine analysis of a large number

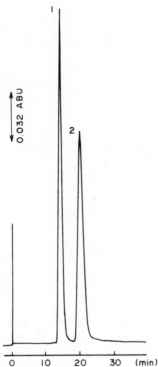

**FIGURE 7-55.** Separation of the optical isomers of *D*- and *L*-binaphthol. Column: μS-Chiralpak OT(+). Column size: 250-mm × 1.5-mm ID. Mobile phase: methanol. Flow rate: 40 μ l/min. Detection wavelength: 265nm. Peaks: 1, *D*-binaphthol; 2, *L*-binaphthol.

of samples. Analysis time in HPLC could be reduced by increasing the flow rate of the mobile phase or reducing the column length. However, this is meaningless if it results in an increase of the operating costs and poor sample separation. The important point in high-speed HPLC is to reduce the analysis time without impairing the separation. It is also important to decide to what extent the analytical time should be reduced. To establish the extent of time reduction, the processing speed of the peripheral equipment, such as the auto-sampler and data processor, as well as the time required for manual operations, such as sample preparation and data evaluation, should be taken into consideration. From a practical point of view, the analytical time could be in the range of 1 to 3 minutes, i.e., reduced to about ⅒th, as compared to the time required in conventional HPLC.

### 7.4.2. Requirements of High-Speed HPLC

The following parameters are important to consider if you wish to achieve high-speed HPLC with high performance: column dimensions (length and ID), mobile-phase flow rate, detector response time, and extra-column band broadening. The analysis time depends on the column dimen-

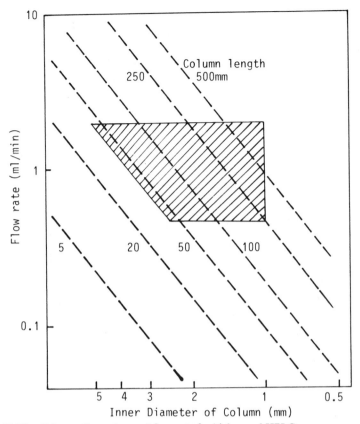

**FIGURE 7-56.** Column dimension and flow rate for high-speed HPLC.

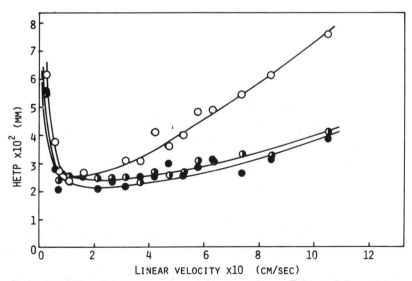

**FIGURE 7-57.** Effect of the detector time constant on column efficiency. Column: 2.1-mm ID × 100-mm packed with Fine SIL C$_{18}$ (5 $\mu$m). Sample: naphthalene (k' = 3.5). Time constants: □, 1.0 sec; ◑, 0.25 sec; ●, 0.1 sec.

A                                    B

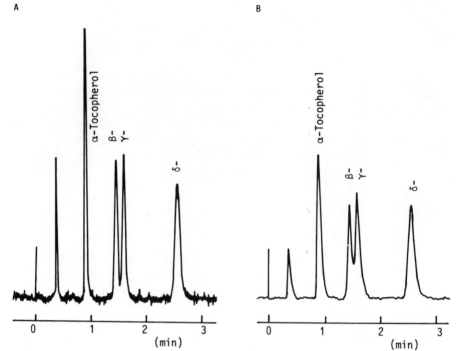

**FIGURE 7-58.** Effect of time constant on resolution. (A) Fast response (time constant: 0.2 sec); (B) slow response (time constant: 2.0 sec). Column: F-Finepak SIL. Column size: 4.6-mm ID × 50-mm, packed with 3 μm silica gel. Mobile phase: n-hexane/isopropanol/acetic acid (100/0.5/0.5). Flow rate: 2.0 ml/min. Fluorescence detector: excitation wavelength, 295 nm; emission wavelength, 325 nm. Volume of flow cell: 6.5 μl. Sample: α-, β-, γ-, and δ-tocopherols. Sample volume: 3 μl.

sions and the mobile-phase flow rate. Their interrelationship in high-speed HPLC is illustrated in Fig. 7-56. The column dimensions and mobile-phase flow rates in the shaded area are recommended, considering the pressure drop, the particle diameter of commercially available packing materials, resolution, and solvent consumption, to give analysis times corresponding to one-tenth of those achieved in conventional HPLC.

High-speed HPLC requires a fast detector response, because each sample component elutes in a very short time. Figure 7-57 shows the relationship between linear velocity of the mobile phase and column efficiency at various detector responses. Slow detector response (time constant: 1.0 sec) causes a decrease in column efficiency at the high linear velocities used in high-speed HPLC. On the other hand, the column efficiency measured by a fast-response detector (time constant, below 0.25 sec) is not decreased seriously, even at high linear velocities. It is clear that the time constant of the detector has to be below 0.25 sec in order to successfully carry out high-speed HPLC. Figure 7-58 shows the effect of the detector response time on peak resolution in an actual chromatogram.

High-speed HPLC also requires low extra-column band broadening because each solute band has a very small dispersion, due to the short column length and small particle diameter of the packing material. Figure 7-59 illustrates the extra-column effects of the injector, connecting tube, and the flow-through cell on resolution in high-speed HPLC using a 50-mm × 4.6-mm ID column. It is clear from this figure that a properly designed system is needed to successfully perform high-speed HPLC.

Figure 7-60 compares the separation of vitamin E isomers obtained by high-speed and conventional HPLC. It can be seen that, while the separation remained almost the same, the analysis time was reduced to about one-tenth of the time needed in conventional HPLC.

### 7.4.3. Analysis of Components Present in Medicines and Cosmetics

HPLC is widely used for the quality control of pharmaceuticals and high-speed HPLC will certainly play an important role here.

Figures 7-61 and 7-62 show the analysis of different commercial cold medicines using columns containing 5 μm particles of a packing consisting of bonded octadecyl groups.

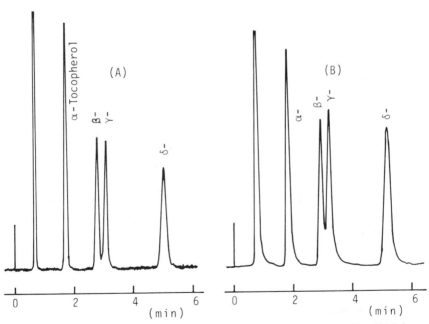

**FIGURE 7-59.** Influence of extra-column band broadening on peak resolution. (A) Injector: Microloop injector (3.0 μl). Flow cell volume: 6.5 μl. Diameter of connecting tube: 0.1 mm. (B) Injector: conventional variable-loop injector (3.0 μl injection). Flow cell volume: 15.0 μl. Diameter of connecting tube: 0.25 mm. Flow rate (A and B): 1.0 ml/min. Other conditions same as in Fig. 7-58.

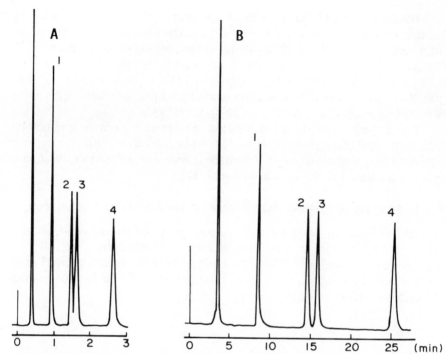

**FIGURE 7-60.** Comparison of (A) high-speed and (B) conventional HPLC. (A) Column: F-Finepak SIL Column size: 4.6-mm ID × 50-cm, packed with 3-μm silica gel. Mobile phase: *n*-hexane/isopropanol/acetic acid (99/1/0.5). Flow rate: 2.0 ml/min. Fluorescence detector with 6.5 μl flow cell: excitation wavelength, 285 nm; emission wavelength, 340 nm. Time constant of the detector: 0.2 sec. (B) Column: Finepak SIL. Column size: 4.6-mm ID × 25-cm packed with 5-μm silica gel. Flow rate: 1.0 ml/min. Time constant of detector: 2.0 sec. Other conditions are the same as in high-speed-HPLC. Peaks: 1, α-tocopherol; 2, β-tocopherol; 3, γ-tocopherol; 4, δ-tocopherol.

In earlier examples, the need to check therapeutic levels of various drugs in serum has already been emphasized. Figure 7-63 illustrates this by showing the analysis of anticonvulsants in serum, using high-speed HPLC.[39]

Two more examples are shown in Figure 7-64, which represents the analysis of tocopherol acetate in a commercial skin cream, and Fig. 7-65, which shows the analysis of tocopherol isomers in a vitamin E capsule. Both represent high-speed analysis.[39]

### 7.4.4. Other Applications

In conventional HPLC, a considerably long time is required for the elution of all the components present. Therefore, gradient elution is usually applied to reduce the analytical time. However, this problem can also be

**FIGURE 7-61.** Separation of components of a commercial cold medicine. Column: 10-cm × 2.1-m ID, packed with Fine SIL $C_{18}$ (5 μm). Mobile phase: acetonitrile/0.02 M sodium 1-pentanesulfonate (50/50). Flow rate: 0.7 ml/min. Detection wavelength: 215 nm. Detector time constant: 0.05 sec.

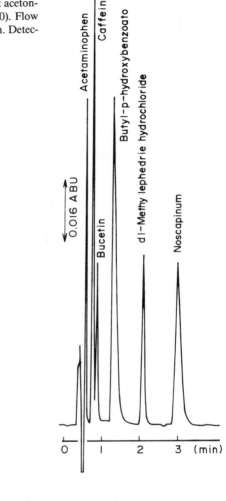

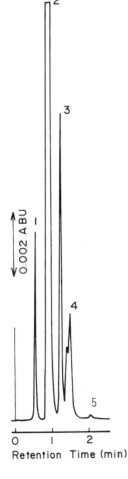

**FIGURE 7-62.** Analysis of components of a commercial cold medicine. Column: F-Finepak SIL $C_{18}$, 5-cm × 4.6-mm ID, packed with 3-μm silica-ODS particles. Mobile phase: methanol/0.4% $KH_2PO_4$/0.4% sodium 1-pentanesulfonate (35/32.5/32.5). Flow rate: 1.0 ml/min. Detection wavelength: 260 nm. Detector time constant: 0.05 sec. Sample: extract of cold medicine using the mobile phase as solvent. Injection volume: 2 μl. Peaks: 1, potassium guaiacolsufonate; 2, acetaminophen; 3, caffeine; 4, riboflavin; 5, *D, L*-methylephedrine hydrochloride.

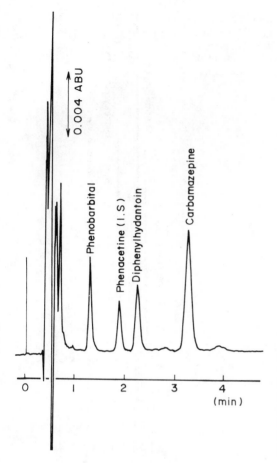

**FIGURE 7-63.** Analysis of anticonvulsants in serum. Column: 2.1-mm ID × 10-cm, packed with Fine SIL $C_{18}T$ (5 μm). Mobile phase: acetonitrile/5mM $KH_2PO_4$ in water (pH: 3.0; 35/65). Flow rate: 0.7 ml/min. Detection wavelength: 220 nm. Detector time constant: 0.05 sec.

solved by using high-speed HPLC with isocratic elution. Figure 7-66 shows the determination of 2, 4-dinitrophenol in polystyrene oligomers by high-speed HPLC.

High-speed HPLC can also be applied with electrochemical detectors. For example, Di Bussolo *et al.* separated and detected phenols, catecholamines, and acetaminophen in 4–6 minutes using 10-cm × 4.6-mm ID and 3.3-cm × 4.6-mm ID columns with an electrochemical detector.[40]

Microbore (1 to 1.5-mm ID) columns can also be used in high-speed HPLC. For example, Scott *et al.*[41] succeeded in separating a seven-component mixture in 30 sec. Kucera, using a 25-cm × 1-mm ID column separated four types of diazepam and their metabolites within 30 sec.[42]

**FIGURE 7-64.** Analysis of tocopherol acetate in a face cream. Column: as in Fig. 7-62. Mobile phase: acetonitrile/0.2% phosphoric acid in water (99/1). Flow rate: 2.0 ml/min. Detection wavelength: 295 nm. Detector time constant: 0.05 sec. Sample: ethanol extract of the face cream. Injection volume: 10 μl. Peak: 1, tocopherol acetate.

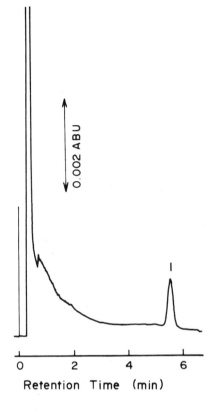

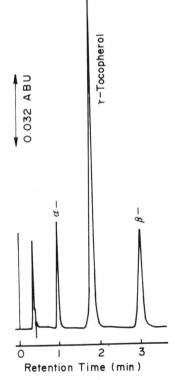

**FIGURE 7-65.** Analysis of α-, β-, and γ-tocopherol isomers in a vitamin E capsule. Column: as in Fig. 7-60. Mobile phase: n-hexane/isopropanol/acetic acid (100/0.4/0.4). Flow rate: 1.0 ml. Detection wavelength: 295 nm. Detector time constant: 0.05 sec. Sample: 2% chloroform solution of the content of the capsule. Injection volume: 2 μl.

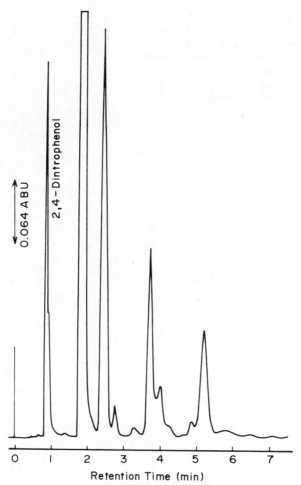

**FIGURE 7-66.** Analysis of 2,4-dinitrophenol in polystyrene oligomers. Column: as in Fig. 7-62. Mobile phase: acetonitrile/0.2% phosphoric acid in water (7/30). Flow rate: 1.0 ml/min. Detection wavelength: 250 nm. Detector time constant: 0.05 sec. Injection volume: 5 μl. Several peaks eluted after 2,4-dinitrophenol correspond to oligostyrene components.

## References

1. Hara, S.; Yamauchi, N.; Nakae, C.; Sakai, S. *Anal. Chem.* **1980, 52,** 33.
2. Kimura, Y.; Kitamura, H.; Araki, T.; Noguchi, K.; Baba, M. *J. Chromatogr.* **1981, 206,** 563.
3. Miyake, K.; Terada, H. *J. Chromatogr.* **1978, 157,** 386.
4. Okuyama, S.; Kokubun, N.; Higashidate, S.; Umemura, D.; Hirata, Y. *Chem. Lett.* **1979,** 1443.
5. Kamada, S.; Maeda, M.; Tsuji, A.; Umezawa, Y.; Kurahashi, T. *J. Chromatogr.* **1982, 239,** 773.
6. Takeuchi, T.; Saito, S.; Ishii, D. *J. Chromatogr.* **1983, 258,** 125.
7. Ishii, D.; Murata, S.; Takeuchi, T. *J. Chromatogr.* **1983, 282,** 569.

8. Takeuchi, T.; Ishii, D. *J. High Resolut. Chromatogr./Chromatogr. Commun.* **1983, 6,** 571.
9. Takeuchi, T.; Yamazaki, M.; Ishii, D. *J. Chromatogr.* **1984, 295,** 333.
10. Takeuchi, T.; Ishii, D.; Mori, S. *J. Chromatogr.* **1983, 257,** 327.
11. Ishii, D.; Takeuchi, T. *J. Chromatogr.* **1983, 255,** 349.
12. Takeuchi, T.; Ishii, D. *J. High Resolut. Chromatogr./Chromatogr. Commun.* **1983, 6,** 310.
13. Takeuchi, T.; Ishii, D. *J. Chromatogr.* **1983, 279,** 439.
14. Takeuchi, T.; Ishii, D. *J. Chromatogr.* **1982, 253,** 41.
15. Takeuchi, T.; Jin, Y.; Ishii, D. *J. Chromatogr.* **1985, 321,** 159.
16. Nakanishi, A.; Ishii, D.; Takeuchi, T. *J. Chromatogr.* **1984, 291,** 398.
17. Ishii, D.; Goto, M.; Takeuchi, T. *J. Chromatogr.* **1984, 316,** 441.
18. Takeuchi, T.; Ishii, D. *J. High Resolut. Chromatogr./Chromatogr. Commun.* **1984, 7,** 151.
19. Iwai, A., Department of Pharmacy, National Toneyama Hospital, Toyonaka-city, Osaka, Japan, personal communication, 1981.
20. Morishita, T., Faculty of Home Economics, Mukogawa Women's University, Hyogo, Japan, personal communications, 1980.
21. Hibi, K.; Kaneuchi, H. In "LC-Family," Japan Spectroscopic Co., Ltd. Press: Tokyo, Japan, 1980; Vol. 15, pp. 14–15.
22. Ouano, A.C. *Ind. Eng. Chem. Fundamentals.* **1972, 11,** 268.
23. Takeuchi, T.; Watanabe, Y.; Matsuoka, K.; Ishii, D. *J. Chromatogr.* **1981, 216,** 153.
24. Takeuchi, T.; Ishii, D. *J. Chromatogr.* **1982, 244,** 23.
25. Klesper, E.; Corwin, A.H.; Turner, D.A. *J. Org. Chem.* **1962, 27,** 700.
26. Takeuchi, T.; Ishii, D.; Saito, M.; Hibi, K. *J. Chromatogr.* **1984, 295,** 323.
27. Yamazaki, M.; Ichinoki, S.; Igarashi, R. *Bunseki Kagaku.* **1981, 30,** 40.
28. White, E.R.; Laufer, D.N. *J. Chromatogr.* **1984, 290,** 187.
29. Baram, G.I.; Grachev, M.A.; Komarova, N.I.; Perelroyzen, M.P.; Bolvanov, Yu.A.; Kuzmin, S.V.; Kargaltsev, V.V. Kuper, E.A. *J. Chromatogr.* **1983, 264,** 69.
30. Takahashi, M. Pittsburgh Conference on Analytical Chemistry and Applied Spectroscopy, Atlantic City, NJ., March 1983, 936.
31. Hibi, K. unpublished results, Japan Spectroscopic Company, Hachioji-city, Tokyo, Japan, 1983.
32. Kucera, P. *J. Chromatogr.* **1980, 198,** 93.
33. Hayes, M.J.; Schwarz, H.E.; Vours, P.; Karger, B.L.; Thurston, A.D. Jr.; McGuire, J.M. *Anal. Chem.* **1984, 56,** 1229.
34. Scott, R.P.; Kucera, P. *J. Chromatogr.* **1979, 185,** 27.
35. Schwarz, H.E.; Karger, B.L.; Kucera, P. *Anal. Chem.* **1983, 55,** 1752.
36. Kucera, P.; Manius, G. *J. Chromatogr.* **1981, 216,** 9.
37. Jin, Z.; Rappaport, S.M. *Anal. Chem.* **1983, 55,** 1778.
38. Okamoto, Y.; Honda, S.; Okamoto, I.; Yuki, H. *J. Am. Chem. Soc.* **1981, 103,** 6971.
39. Hibi, K. In "LC-Family," Japan Spectroscopic Co., Ltd. Press: Tokyo, Japan, 1984, Vol. 19, pp. 10–13.
40. Di Bussolo, J.M.; Dong, M.W.; Gant, J.R. *J. Liquid Chromatogr.* **1983, 6,** 2353.
41. Scott, R.P.W.; Kucera, P.; Munroe, M. *J. Chromatogr.* **1979, 186,** 475.
42. Kucera, P. *J. Chromatogr.* **1980, 198,** 93.

# Appendixes

## List of Available Packing Materials for the Preparation of Packed and Semi-Microcolumns and Microcolumns*

*Adapted from R.E. Majors (*J. Chromatogr. Sci.,* 18, 488–511 (1980). Reprinted with some modifications with the permission of Preston Technical Abstracts Company.

## APPENDIX 1

### Microparticulate Packings for Liquid-Solid Chromatography

| Type | Name | Supplier(s) | Average particle diameter ($\mu$m) | Specific surface area ($m^2/g$) | Average pore diameters (Å) | Description and further specifications |
|---|---|---|---|---|---|---|
| Silica, irregular | Chromosorb LC-6 | 1, 41 | 5, 10 | 400 | 120 | Density: 0.40 g/ml |
| | Hitachi Gel 3030 series | 22 | 5–7 | 500 | | |
| | ICN Silica | 24 | 3–7, 7–12 | 500–600 | | |
| | Lichrosorb Si-60 | 2, 5, 10, 12, 13, 23, 27, 36, 41, 44 | 5, 10 | 550 | 60 | |
| | Lichrosorb Si-100 | 2, 5, 10, 12, 13, 23, 27, 36, 41, 44 | 5, 10 | 300 | 100 | |
| | $\mu$ Porasil | 1, 46 | 10 | 300–350 | | |
| | Partisil | 2, 18, 23, 27, 33, 47 | 5, 10 | 400 | 60 | Density: 0.45 g/ml |
| | Polygosil 60 | 14, 23, 29, 34, 38 | 5, 7, 10, 15 | 500 | 60 | Pore volume; 0.75 ml/g |
| | R Sil | 1, 38 | 5, 10 | 550 | 60 | |
| | Sil-X-1 | 31 | 13 ± 5 | 400 | | Chemically-treated surface |
| | Apex Silica | 27 | 5 | 200 | 100 | Pore volume: 0.7 ml/g |
| | Fine SIL | 25 | 5 | 500 | 60 | |
| | Hypersil | 12, 23, 27 | 3, 5, 10 | 200 | 100 | |

| | Suppliers[a] | | | | Notes |
|---|---|---|---|---|---|
| **Silica, spherical** | Lichrosphere Si-100 | 1, 5, 21, 23, 27, 41 | 5, 10 | 250 | 100 | Pore volume: 1.2 mg/g, Void volume: 1.06 ml/g, Different pore diameters available |
| | Nucleosil Si-50 | 1, 14, 23, 29, 34, 38, 41 | 5, 7.5, 10 | 300, 500 | 50 | Pore volume: 0.8 ml/g |
| | Nucleosil Si-100 | 1, 14, 23, 29, 38, 41 | 5, 7.5, 10 | 300 | 100 | Pore volume: 1.0 ml/g |
| | Separon Si VSK | 28 | 5, 7.5, 10 | 450 | 130 | Pore volume: 1.5 ml/g |
| | Spherisorb SW | 1, 12, 20, 23, 27, 32, 35, 41 | 3, 5, 10 | 220 | 80 | Packing density: 0.6 ml/g |
| | Spherosil XOA 600 | 3, 37, 41 | 5–7 | 600 ± 10% | 80 | Pore volume: 0.7–1 ml/g |
| | Spherosil XOA 800 | 3, 37, 41 | 5–7 | 860 | 40 | Pore volume: 0.4–0.6 ml/g |
| | TSK gel Silica-60 | 43 | 5, 10 | 500 | 60 | |
| | TSK gel Silica-150 | 43 | 5, 10 | 330 | 150 | |
| | Vyda 101 TP | 1, 3, 4, 5, 14, 41 | 10 | 100 | 330 | |
| | Zorbax BP-SIL | 1, 16, 23, 41 | 8 | 350 | 70–80 | |
| **Alumina** | Alox 60-D | 14, 29, 34, 38 | 5, 10 | 60 | 60 | Basic: pH = 9.5 |
| | Lichrosorb Alox T | 10, 13, 35, 36 | 5, 10 | 70 | 150 | |
| | Spherisorb AT | 23, 32, 35, 41 | 5, 10 | 95 | 130 | pH Limit: 10 |

[a]See Appendix 10 for list of suppliers.

## APPENDIX 2

### Bonded-phase Microparticulate Packings for Normal-phase Chromatography[a]

| Name | Supplier(s)[b] | Functionality | Average particle diameter ($\mu$m) | Description and further specifications |
|---|---|---|---|---|
| Lichrosorb Diol | 2, 5, 8, 10, 12, 13, 21, 23, 27, 36, 44 | Diol | 10 | For very polar compounds |
| Nucleosil OH | 1, 2, 14, 29, 34, 38 | Alcoholic OH | 7.5 | Wettable with water |
| TSK gel OH-120 | 43 | Alcoholic OH | 5, 10 | |
| Nucleosil N(CH$_3$)$_2$ | 1, 2, 14, 23, 29, 34, 38 | Trialkylamine | 5, 10 | Also weak anion exchanger<br>Weakly basic<br>For separation of weak anion phenols |
| Polygosil 60-D-N(CH$_3$)$_2$ | 1, 23 | Trialkylamine | 5, 10 | Can also be used as a weak anion exchanger |
| Polygosil NO$_2$ | 1, 2, 14, 23, 29, 34, 38, 41 | Nitro | 5, 10 | Spherical<br>Affinity for double bonds |
| Polygosil NO$_2$ | 1, 23, 29 | Nitro | 5, 10 | For separation of aromatics<br>Affinity for double bonds |
| RSIL NO$_2$ | 1, 38 | Nitro | 5, 10 | Coverage: 5% C |

[a]Weakly polar.
[b]See Appendix 10 for list of suppliers.

Cyano Bonded-phase Microparticulate Packings for Normal-phase Chromatography[a]

| Name | Functionality | Average particle diameter ($\mu$m) | Supplier(s)[b] | Description and further specifications |
|---|---|---|---|---|
| Apex Cyano | Cyano | 5 | 27 | Monolayer |
| Chromosorb LC-8 | Cyano | 5, 10 | 1, 8, 45 | Pore diameter: 110Å, Density: 0.43 g/ml |
| CPS Hypersil | Cyanopropyl | 3, 5, 10 | 23, 41 | pH: 3–8 Monolayer coverage |
| Lichrosorb CN | Cyano | 5, 10 | 5, 10, 13, 23, 30, 36 | Prepared by first bonding tolyl-trichloro silane followed by NBS treatment and nucleophilic displacement of halogen |
| $\mu$ Bondapak CN | Cyano | 10 | 1, 46 | 9 wt% loading |
| Micropak CN | Cyanopropyl | 10 | 45 | |
| Nucleosil CN | Cyano | 5, 10 | 1, 2, 14, 29, 34, 38, 41 | Coverage: 6 $\mu$eq/m$^2$ |
| Partisil-10 PAC | Cyano amino | 10 | 1, 2, 12, 18, 23, 27, 33, 47 | 2:1 amino to cyano ratio |
| Polygosil CN | Cyano | 5, 10 | 1, 23, 29 | Coverage: 6 $\mu$eq/m$^2$ |
| RSIL CN | Cyanopropyl | 5, 10 | 1, 38 | 5% C loading |
| Spherisorb CN | Cyano | 5, 10 | 1, 2, 12, 20, 23, 27, 32, 35, 41 | Pore diameter: 80Å Coverage: 0.6 mmole/g |
| Vydac 501TP | Cyano | 10 | 1, 3, 4, 5, 14, 39, 41 | Spherical |
| Zorbax BP-CN | Cyano | 8 | 8, 16, 23, 41 | Monolayer coverage |

[a]Moderately polar.
[b]See Appendix 10 for list of suppliers.

Amino Bonded-phase Microparticulate Packings for Normal-phase Chromatography[a]

| Name | Supplier(s)[b] | Functionality | Average particle diameter ($\mu$m) | Description and further specifications |
|---|---|---|---|---|
| Apex Amino | 27 | Amino | 5 | |
| APS Hypersil | 23, 32 | Aminopropyl | 3, 5, 10 | |
| Chromosorb LC-9 | 26, 41 | Amino | 10 | Density: 0.48 g/ml |
| Fine SIL NH$_2$ | 25 | Aminopropyl | 5, 10 | Spherical |
| Lichrosorb NH$_2$ | 1, 2, 5, 8, 13, 21, 23, 27, 44 | Amino | 10 | |
| $\mu$ Bondapak NH$_2$ | 1, 23, 46 | Amino | 10 | pH: 2–8 Coverage: 9 wt % |
| Micropak NH$_2$ | 45 | Aminopropyl | 10 | |
| Nucleosil NH$_2$ | 1, 2, 14, 23, 29, 34, 38, 41 | Amino | 5, 10 | Weakly basic For the separation of polar compounds |
| Polygosil NH$_2$ | 1, 23, 29 | Amino | 5, 10 | |
| RSIL NH$_2$ | 1, 38 | Aminopropyl | 5, 10 | Prepared by first bonding chloropropyltrichlorosilane, followed by NBS treatment and nucleophilic displacement of halogen 6% loading |
| Separon Si NH$_2$ | 28 | Aminopropyl | 5, 10 | |
| Spherisorb NH$_2$ | 1, 12, 20, 23, 27, 32, 35, 41 | Aminopropyl | 5, 10 | Pore diameter: 80Å Coverage: 0.6 mmole/g |
| TSK gel NH$_2$-60 | 43 | Aminopropyl | 5 | |
| Zorbax BP-NH$_2$ | 8, 23, 41 | Amino | 8 | Monolayer |

[a]Highly polar.
[b]See Appendix 10 for list of suppliers.

## APPENDIX 5

### Octadecylsilane Bonded-phase Microparticulate Packings for Reversed-phase Chromatography

| Name | Supplier(s)[a] | Average particle diameter ($\mu$m) | Coverage | Description and further specifications |
|---|---|---|---|---|
| Apex ODS | 27 | 5 | 10% C | Endcapped |
| Chromosorb LC-7 | 1, 41 | 3, 5, 10 | 15% C | Monolayer<br>Average pore diameter: 100Å |
| Fine SIL C$_{18}$ | 25 | 5, 10 | 14% C | Encapped |
| Fine SILC C$_{18}$T | 25 | 5, 10 | 14% C | pH range: 2–8 |
| Hitachi Gel 3050 series | 22 | 5–7, 10–15 | | pH range: 1–9 |
| Lichrosorb RP-18 | 1, 5, 10, 12, 13, 21, 23, 27, 36, 41, 44 | 5, 10 | 22% C | |
| $\mu$ Bondapak C$_{18}$ | 1, 23, 46 | 10 | 10% C | Polymeric |
| Micro Pak CH | 45 | 10 | 22% C | Encapped<br>Recommended for nonpolar samples |
| Micro Pak MCH | 45 | 5, 10 | 12% C | Monomeric layer<br>Available in endcapped and nonendcapped versions |
| Nucleosil C$_{18}$ | 1, 14, 23, 29, 34, 38 | 5, 7.5, 10 | 15–16% C | Capacity factor double of C$_8$<br>pH range: 1–9<br>Spherical |
| ODS Hypersil | 12, 23, 27, 40 | 3, 5, 10 | 9% C | Endcapped |
| ODS-Sil-X-1 | 31 | $13 \pm 5$ | | |
| Partisil ODS-1 | 1, 18, 23, 27, 33, 47 | 5, 10 | 5% C | For more polar solutes<br>High silanol content |
| Partisil-10 ODS-2 | 1, 12, 18, 23, 33, 47 | 10 | 15% | High retention and loading capacity<br>Temperature to 70°C |

| Name | Supplier(s)[a] | Average particle diameter ($\mu$m) | Coverage | Description and further specifications |
|---|---|---|---|---|
| Partisil-10 ODS-3 | 18, 33, 47 | 10 | 10% C | Endcapped |
| Polygosil $C_{18}$ | 1, 14, 23, 24, 29, 34, 38 | 3, 7.5, 10 | 11% C | Capacity factor: double of $C_8$ pH range: 1–9 irregular |
| RSIL $C_{18}$ HL | 1, 38 | 5, 10 | 18% C | Encapped |
| RSIL $C_{18}$ LL | 1, 38 | 5, 10 | 9% C | Endcapped |
| Separon Si $C_{18}$ | 28 | 5, 10 | 20% C | Endcapped |
| Spherisorb ODS | 1, 12, 20, 23, 27, 32, 35, 41 | 5, 10 | 7% C | Endcapped |
| | | | 0.3 mmole/g | Average pore diameter: 80Å |
| Spherosil $C_{18}$ | 3, 31, 37, 41 | 5–7 | 20–23% C | |
| Techsphere $C_{18}$ | 23 | 5, 10 | 10% C | Irregular version called Techsil $C_{18}$ Both available in endcapped and nonendcapped versions |
| TSK gel ODS 120A | 43 | 5, 10 | | |
| TSK gel ODS 120T | 43 | 5, 10 | | |
| Vydac 201 $C_{18}$ | 1, 3, 4, 5, 14, 39, 41 | 5, 10 | 10% 3.35 $\mu$mole/m$^2$ | Nonendcapped pH range: 1–9 |
| Zorbax BP-ODS | 1, 12, 16, 23, 41 | 8 | 15% C | Monolayer |

[a]See Appendix 10 for list of suppliers.

## APPENDIX 6

### Octylsilane Bonded-phase Microparticulate Packings for Reversed-phase Chromatography

| Name | Supplier(s)[a] | Average particle diameters ($\mu$m) | Coverage | Description and further specifications |
|---|---|---|---|---|
| Apex C$_8$ | 27 | 5 | 5% | Endcapped |
| Fine SIL C$_8$ | 25 | 5, 10 | | |
| Lichrosorb RP-8 | 1, 5, 10, 12, 13, 21, 23, 27, 36, 41, 44 | 5, 10 | 13–14% C | Recommended for samples of moderate polarity |
| MOS Hypersil | 12, 23, 40 | 3, 5, 10 | | Monolayer |
| Nucleosil C$_8$ | 1, 14, 23, 29, 34, 38, 41 | 5, 7.5, 10 | 10–11% C | pH range: 1–9 |
| Polygosil C$_8$ | 14, 23, 29, 36, 38 | 5, 7.5, 10 | 10–11% C | Spherical pH range: 1–9 |
| Techsphere C$_8$ | 23 | 5, 10 | 10% C | Irregularly shaped version called Techsil C$_8$ |
| Zorbax BP-C$_8$ | 12, 16, 23, 41 | 8 | 15% C | Monolayer |

[a]See Appendix 10 for list of suppliers.

**APPENDIX 7**

Other Microparticulate Packings for Reversed-phase Chromatography

| Type | Name | Supplier(s)[a] | Average particle diameter ($\mu$m) | Coverage | Description and further specifications |
|---|---|---|---|---|---|
| Hexyl-silane | Chromegabond cyclohexyl | 17 | 10 | 10% C | |
| | | 17 | 10 | 10% C | |
| | Spherisorb C$_6$ | 1, 12, 20, 23, 33, 35, 41 | 5 ± 2 | 0.6 mmole/g | Monolayer
Average pore diameter: 80Å
Endcapped |
| | Apex C$_2$ | 27 | 5 | | For polar multifunctional solutes |
| Methyl-silane | Fine SIL C$_1$ | 25 | 5, 10 | | |
| | Lichrosorb RP-2 | 1, 5, 8, 10, 12, 13, 21, 23, 27, 36, 41, 44 | 5, 10 | 3% C | Recommended for polar compounds |
| | RSIL C$_3$ | 1, 38 | 5, 10 | 7% C | Chloropropyltrichlorosilane reactant + TMCS endcapping |
| | Separon SiC$_1$ | 28 | 5, 10 | | Methylated surface useful in exclusion chromatography
Exclusion limit: 50,000 |
| | TSK gel TMS-250 | 43 | 5 | | Monolayer |
| | Zorbax BP-TMS | 12, 16, 23 | 8 | | |

| | Material | | Particle size | Carbon load | Comments |
|---|---|---|---|---|---|
| Phenyl-silane | Apex Phenyl | 27 | 5 | 5% C | For amines and hydroxyl compounds; More polar than C$_{18}$ |
| | μ Bondapak phenyl | 1, 23, 46 | 10 | 10% C | For very nonpolar compounds and fatty acids |
| | Nucleosil Phenyl | 1, 14, 29, 34, 38 | 7.5 | 10% C | Recommended for nonpolar samples |
| | RSIL Phenyl | 1, 38 | 5, 10 | | |
| | Spherisorb P | 1, 12, 23, 32 | 5 ± 2 | 0.3 mmole/g | Average pore diameter: 80Å |
| Styrene-divinyl benzene copolymer | Fine GEL 110 | 25 | 10 | | Useful on SEC; Exclusion limit: 3000 |
| | Fine GEL 101 | 25 | 8 | | |
| | Hitachi GEL 3011 | 22 | 10–15 | | Spherical pH range: 2–11; Functionality: CH$_2$OH; Can be used in size-exclusion chromatography |
| | Benson BN | 7 | 7–10 | | Available in different crosslinkings (4, 7, 8 and 10% divinylbenzene) |

Microparticulate Porous Packings for Anion-exchange Chromatography

| Type | Name | Supplier(s)[a] | Average particle diameter ($\mu$m) | Strength[b] | Functional group | Ion-exchange capacity (meq/g) | Description and further specifications |
|------|------|-----------|---------|----------|------------------|------------------|------------------|
| | Lichrosorb AN | 5, 8, 10, 13, 27, 36 | 10 | S | $-NR_3^+$ | 0.55 | On 100 Å silica |
| | Micropak AX | 45 | 5, 10 | W | Difunctional amine | | Recommended for the analysis of nucleic acid constituents |
| | Micropak SAX | 45 | 10 | S | $-NR_3^+$ | | General purpose anion exchange |
| | | | | | | | Also for nucleotides |
| | Nucleosil SB | 1, 2, 10, 14, 23, 34, 38 | 5, 10 | S | $-NMe_3^+ \ Cl^-$ | 1 | pH range: 1–9 |
| | Partisil 10SAX | 2, 18, 23, 27, 33, 47 | 10 | S | $-NR_3^+$ | <1 (estimated) | pH range: 1.5–7.5 |
| | RSIL AN | 1, 38 | 5, 10 | S | $-NR_3^+$ | <1 (estimated) | |
| | Synchropak AX-300 | 4, 5, 35, 42, 45 | 10 | W | Polymeric amine | | Average pore diameter: 300Å |
| | | | | | | | pH range: 2–8 |
| | | | | | | | Recommended for proteins and enzymes |
| | | | | | | | Monolayer coverage |

|  | Name | Suppliers[a] |  | S/W[b] | Functional group |  | Comments |
|---|---|---|---|---|---|---|---|
|  | Vydac 301 TP | 1, 3, 4, 5, 14, 39, 41 | 10 | S | $-NR_3^+$ | 0.2 | pH range: 1–9<br>High-density hydrophobic phase protects silica |
| Resin based | Zorbax SAX | 16, 23 | 7 | S | $-NR_3^+$ | <1 (estimated) | pH range: 2–9 |
|  | Aminox<br>A-28<br>A-29 | 9, 45 | $9 \pm 2$<br>$7 \pm 10$ | S | $-NR_3^+$ | 3.2 | All are 8% crosslinked |
|  | Benson BA-X | 1, 7, 11 | 7–10 | S | $-NR_3^+Cl^-$ | 5 | pH range: 1–12<br>Crosslinking (×4, ×6, ×8, ×10)<br>10–15 meq/g also available |
|  | Benson BWA | 7 | 7–10 | W | $-NR_2HCl$ | 5 | pH range: 1–12 |
|  | Hitachi GEL 3011-N | 22 | 10–15 | S | $-NR_3^+$ |  | pH range: 2–11<br>Spherical |

[a]See Appendix 10 for list of suppliers.
[b]S, strong; W, weak.

## APPENDIX 9

### Microparticulate Porous Packings for Cation-exchange Chromatography

| Type | Name | Supplier(s)[a] | Average particle diameter (μm) | Strength[b] | Functional group | Ion-exchange capacity (meq/g) | Description and further specifications |
|---|---|---|---|---|---|---|---|
| Silica based | Chromegabond SCX | 6 | 10 | S | $-SO_3H$ | <1 (estimated) | Average pore diameter: 60 Å |
| | Lichrosorb KAT | 1, 5, 8, 10, 13, 23, 36 | 10 | S | $-SO_3^-$ | 1, 2 | On 100Å Silica |
| | Nucleosil SA | 1, 2, 14, 23, 29, 34, 38 | 5, 10 | S | $-SO_3^-Na^+$ | ~1 | pH range: 1–9 |
| | Partisil-10 SCX | 1, 2, 18, 23, 27, 33, 47 | 10 | S | $-SO_3^-$ | ~1 (estimated) | pH range: 1.5–7.5 |
| | RSIL-CAT | 1, 38 | 5, 10 | S | $-SO_3^-$ | ~1 | 5% loading |
| | Vydac 401 TP | 1, 3, 4, 14, 15, 39, 41 | 10 | S | $-SO_3^-$ | ~1 (estimated) | |
| | Zorbax SCX | 16, 23 | 6–8 | S | $-SO_3^-$ | 5 | pH range: 2–9 |
| | Aminex A series | 9, 45 | | S | $-SO_3^-$ | 5 | All 8% crosslinked |
| | A-5 | | 13 ± 2 | | | | |
| | A-7 | | 7–11 | | | | |
| | A-8 | | 5–8 | | | | |
| | A-9 | | 11.5 ± 0.5 | | | | |
| | Aminex HPX-87 | 9 | 9 | S | $-SO_3^-$ | 5 | |

| | | Ref.[a] | Mesh size | Type[b] | Functional group | pK | Comments |
|---|---|---|---|---|---|---|---|
| Resin based | Beckman AA series | 6 | | S | $-SO_3^-$ | 5 | All are 8% crosslinked |
| | W-1 | | $12.0 \pm 2.5$ | | | | Used mainly for amino acid analysis |
| | W-2 | | $9.5 \pm 2.0$ | | | | |
| | W-3 | | $8.5 \pm 2.5$ | | | | |
| | W-4 | | $8.5 \pm 1.0$ | | | | |
| | Benson BCOOH | 7 | 7–10 | W | $-COOH$ | 10 | |
| | Benson BC-X | 1, 7, 11 | 7–10 | S | $-SO_3^-Na^+$ | 5.2 | pH range: 1–14 |
| | | | 10–15 | | | | Crosslinking ($\times 4$ to $\times 32$) |
| | | | | | | | Can also be used in reversed-phase chromatography |
| | Chromex Cation | 15 | $11 \pm 1$ | S | $-SO_3^-$ | 5 | Crosslinking (8 or 12%) General purpose |
| | Hamilton HC | 19 | 10–15 | S | $-SO_3^-Na^+$ | 5 | |
| | Hitachi GEL 3011C | 22 | 10–15 | W | $-COOH$ | | |
| | Hitachi GEL 3011-S | 22 | 10–15 | S | $-SO_3^-$ | 5 | pH range: 2–11 Spherical |

[a]See Appendix 10 for list of suppliers.
[b]S, strong; W, weak.

## APPENDIX 10
### List of Suppliers Cited in Appendixes 1 to 9

1. Alltech
2. Altex
3. Analabs
4. Anspec Company
5. Applied Science
6. Beckman
7. Benson Company
8. Biolab Products
9. BioRad Laboratories
10. Brownlee
11. Calbiochem-Behring Corporation
12. Chromanetics
13. Chromatix
14. Chrompak (Holland)
15. Dionex
16. DuPont
17. ES Industries
18. Gow Mac
19. Hamilton
20. Hetp
21. Hewlett-Packard
22. Hitachi
23. HPLC Technology Ltd.
24. ICN Inc.
25. Jasco
26. Johns-Manville
27. Jones Chromatography, Inc.
28. Laboratory Instrument Works (Prague, Czechoslovakia)
29. Machery-Nagel (Germany)
30. MCB
31. Perkin-Elmer
32. Phase Sep
33. Pierce
34. Rainen
35. Regis
36. Rheodyne
37. Rhone Poulenc (France)
38. RSL
39. Separations Group
40. Shandon
41. Supelco
42. Synchrom
43. Toyo Soda
44. Unimetrics Corporation
45. Varian
46. Waters Associates
47. Whatman

# List of Symbols

| | |
|---|---|
| $A$ | Peak area. A constant depending on the reactor-bed geometry |
| $AU$ | UV detector output in terms of absorbance units |
| $AU_{p(max)}$ | Maximum absorbance of the peak |
| $C$ | Capacitance or concentration |
| $C_f$ | Concentration of the final solvent in gradient elution |
| $C_i$ | Concentration of the initial solvent in gradient elution |
| $C_{max}$ | Maximum peak concentration |
| $C_M$ | Term expressing molecular diffusion in the mobile phase |
| $C_{sol}$ | Solute concentration |
| $C_s$ | Term expressing molecular diffusion in the stationary phase |
| $C_t$ | Concentration of the mixed solvent in gradient elution at time $t$ |
| $d_c$ | Column diameter or coil diameter of the open-tubular reactor |
| $d_{dc}$ | Detector-cell diameter |
| $d_f$ | Thickness of the stationary-phase film |
| $d_p$ | Particle diameter |
| $d_t$ | Internal diameter of the open-tubular reactor |
| $D_M$ | Coefficient of molecular diffusion of the solute in the mobile phase |
| $Dn$ | Dean number |
| $D_s$ | Coefficient of molecular diffusion of the solute in the stationary phase |
| $e$ | Semiderivative of the current with respect of time, in voltammetric detection |
| $E$ | Separation impedance or electrode potential |
| $E_i$ | Electrical input |
| $E_0$ | Electrical output |

| | |
|---|---|
| $f_R$ | Detector response factor for the sample solute |
| $F$ | Volumetric flow rate of the mobile phase |
| | |
| $h$ | Reduced plate height |
| $H$ | Height equivalent to one theoretical plate (HETP) |
| | |
| $k'$ | Capacity factor |
| $k_0$ | Permeability constant of the packing material |
| | |
| $l_{dc}$ | Detector cell pathlength |
| $l_t$ | Tube length |
| $l_{t(max)}$ | Maximum length of the connecting tube |
| $L$ | Column length or length of the open-tubular reactor |
| | |
| $m$ | Semi-integral of the current with respect of time, in voltammetric detection |
| $m_s$ | Sample mass (weight) |
| $MW$ | Gram-molecular weight |
| | |
| $N$ | Number of theoretical plates |
| | |
| $\Delta P$ | Pressure drop along a column or a tube |
| | |
| $r$ | Distance from the tube center in the calculation of Poiseuille distribution |
| $r_0$ | Internal radius of the tube in the calculation of Poiseuille distribution |
| $R$ | Electrical resistance |
| | |
| $Sc$ | Schmit number |
| $S_t$ | Cross-sectional area of the connecting tube |
| | |
| $t_0$ | Elution time of an unretained solute |
| $t$ | Mean residence time in the reactor |
| | |
| $u$ | Linear velocity of the mobile phase |
| $u_{av}$ | Average velocity in Poiseuille flow |
| $U_{max}$ | Maximum velocity in Poiseuille flow |
| $u(r)$ | Flow-velocity profile function in Poiseuille flow |
| | |
| $V_d$ | Detector-cell volume |
| $V_{d(max)}$ | Maximum detector-cell volume |

| $V_{ex}$ | Extracolumn peak volume |
|---|---|
| $V_{inj}$ | Injection volume |
| $V_m$ | Mixing chamber volume |
| $V_0$ | Elution volume of an unretained peak (vacant volume of the column) |
| $V_p$ | Peak volume |
| $V_{p(ob)}$ | Observed peak volume |
| $V_{pt}$ | Peak volume contribution of the tubing due to Poiseuille flow dispersion |
| $V_{res}$ | Volume of solvent reservoir |
| $V_R$ | Retention volume |

## Greek Symbols

| $\alpha$ | alpha; relative retention or separation factor |
|---|---|
| $\gamma$ | gamma; tortuosity factor of the packed-bed reactor |
| $\epsilon$ | epsilon; column porosity |
| $\epsilon_\lambda$ | epsilon; molar absorptivity at a specified wavelength $\lambda$ |
| $\eta$ | eta; viscosity of mobile-phase solvent or volumetric efficiency of reciprocating pump |
| $\nu$ | nu; reduced mobile-phase velocity |
| $\kappa$ | kappa; constant related to the coiling of the open-tubular reactor |
| $\lambda$ | lambda; wavelength |
| $\phi$ | phi; column resistance parameter |
| $\rho$ | rho; density of the fluid in the open-tubular reactor |
| $\sigma$ | sigma; standard deviation of the solute concentration distribution of a peak |
| $\sigma_d^2$ | Variance of a peak due to the contribution of the detector cell |
| $\sigma_{ex}^2$ | Variance of extracolumn peak broadening |
| $\sigma_p^2$ | Variance of a peak corresponding only to the contribution of the column |
| $\sigma_{p(ob)}^2$ | Variance of an observed peak |
| $\sigma_{pt}^2$ | Contribution of Poiseuille flow dispersion in long, straight tubing to the peak variance |
| $\sigma_s^2$ | Variance of sample peak due only to injector contribution |
| $\sigma_t^2$ | Peak variance in terms of time |
| $\sigma_{tc}^2$ | Peak variance by column contribution in terms of time |

| | |
|---|---|
| $\sigma^2_{t(ob)}$ | Observed peak variance in terms of time |
| $\sigma^2_{tr}$ | Peak variance by reactor contribution in terms of time |
| $\sigma^2_v$ | Peak variance in terms of the volume of the mobile phase |
| $\tau$ | tau; detector time constant |
| $\tau_{(max)}$ | maximum time constant |

# Index